D0153568

The Dolciani Mathematical Expositions

NUMBER TWENTY-ONE

Logic as Algebra

Paul Halmos
Santa Clara University

Steven Givant
Mills College

DOWNS-JONES LIBRARY
HUSTON-TILLOTSON COLLEGE

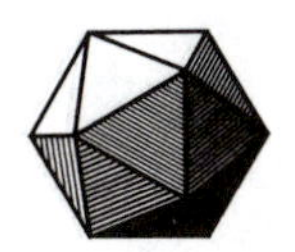

Published and Distributed by
THE MATHEMATICAL ASSOCIATION OF AMERICA

QA
9
.H293
1998

THE
DOLCIANI MATHEMATICAL EXPOSITIONS

Published by
THE MATHEMATICAL ASSOCIATION OF AMERICA

Committee on Publications
JAMES W. DANIEL, *Chair*

Dolciani Mathematical Expositions Editorial Board
BRUCE P. PALKA, Editor
EDWARD J. BARBEAU
IRL C. BIVENS
SUSAN C. GELLER

The DOLCIANI MATHEMATICAL EXPOSITIONS series of the Mathematical Association of America was established through a generous gift to the Association from Mary P. Dolciani, Professor of Mathematics at Hunter College of the City University of New York. In making the gift, Professor Dolciani, herself an exceptionally talented and successful expositor of mathematics, had the purpose of furthering the ideal of excellence in mathematical exposition.

The Association, for its part, was delighted to accept the gracious gesture initiating the revolving fund for this series from one who has served the Association with distinction, both as a member of the Committee on Publications and as a member of the Board of Governors. It was with genuine pleasure that the Board chose to name the series in her honor.

The books in the series are selected for their lucid expository style and stimulating mathematical content. Typically, they contain an ample supply of exercises, many with accompanying solutions. They are intended to be sufficiently elementary for the undergraduate and even the mathematically inclined high-school student to understand and enjoy, but also to be interesting and sometimes challenging to the more advanced mathematician.

DOLCIANI MATHEMATICAL EXPOSITIONS

1. *Mathematical Gems,* Ross Honsberger
2. *Mathematical Gems II,* Ross Honsberger
3. *Mathematical Morsels,* Ross Honsberger
4. *Mathematical Plums,* Ross Honsberger (ed.)
5. *Great Moments in Mathematics* (*Before 1650*), Howard Eves
6. *Maxima and Minima without Calculus,* Ivan Niven
7. *Great Moments in Mathematics* (*After 1650*), Howard Eves
8. *Map Coloring, Polyhedra, and the Four-Color Problem,* David Barnette
9. *Mathematical Gems III,* Ross Honsberger
10. *More Mathematical Morsels,* Ross Honsberger
11. *Old and New Unsolved Problems in Plane Geometry and Number Theory,* Victor Klee and Stan Wagon
12. *Problems for Mathematicians, Young and Old,* Paul R. Halmos
13. *Excursions in Calculus: An Interplay of the Continuous and the Discrete,* Robert M. Young
14. *The Wohascum County Problem Book,* George T. Gilbert, Mark Krusemeyer, and Loren C. Larson
15. *Lion Hunting and Other Mathematical Pursuits: A Collection of Mathematics, Verse, and Stories by Ralph P. Boas, Jr.,* edited by Gerald L. Alexanderson and Dale H. Mugler

16. *Linear Algebra Problem Book,* Paul R. Halmos
17. *From Erdős to Kiev: Problems of Olympiad Caliber,* Ross Honsberger
18. *Which Way Did the Bicycle Go?... and Other Intriguing Mathematical Mysteries,* Joseph D. E. Konhauser, Dan Velleman, and Stan Wagon
19. *In Pólya's Footsteps: Miscellaneous Problems and Essays,* Ross Honsberger
20. *Diophantus and Diophantine Equations,* I. G. Bashmakova
21. *Logic as Algebra,* Paul Halmos and Steven Givant

Contents

Preface

This book is based on the notes of a course in logic that the first author has given several times. We have tried to retain the spirit and purpose of those original notes. What was that purpose? It was to show that logic can (and perhaps should) be viewed from an algebraic perspective. When so viewed, many of its principal notions are seen to be old friends, familiar algebraic notions that were "disguised" in logical clothing. Moreover, the connection between the principal theorems of the subject and well-known theorems in algebra becomes clearer. Even the proofs often gain in simplicity.

We shall treat the propositional calculus and, in addition, the monadic predicate calculus, that is to say, predicate logic with a single quantifier. This allows us to give an algebraic treatment of one of the oldest branches of the subject, the theory of syllogisms, a branch that is neglected in most modern mathematical treatises of logic. An algebraic treatment of all of first-order predicate logic leads naturally to the theory of polyadic algebras, a somewhat more advanced subject.

Our presentation is aimed at a broad audience, from "amateurs" (in the original sense of the word) of mathematics with a potential interest in logic, to undergraduate students, to professional mathematicians, computer scientists, and philosophers. All that is required is a working knowledge of the basic mathematical notions that are studied in a first course in abstract algebra. In particular, the reader is not required to have any previous acquaintance with logic.

What is logic?

1. To count or to think

The dictionary says that the word "geometry" referred originally to land measure. Nowadays, however, no student of geometry approaches the subject in the hope that it will teach him to recapture a plot of ground whose boundaries were obscured by the flooding of the Nile, and no mathematical purist is repelled from the subject by the fear that it will involve him in ad hoc deductions and laborious computations. The content of geometry has become pure mathematics, and the methods of geometry have become purely mathematical.

The dictionary says also that "logic" once had to do with words and with reason, and apparently it still does. Many students approach the study of logic hoping to get philosophical insight into the mysteries of the laws of thought, and many mathematicians are uncomfortable in the presence of what they regard as the pedantic and artificial techniques of the subject that calls itself metamathematics. The content of logic appears to be sentences and deductions, and the methods of logic appear to be enumerative (counting) and combinatorial (arranging). Classical analysts and homological algebraists view such goings on with suspicion, and their admiration of an occasional spectacular breakthrough in logic is grudging and puzzled.

Combinatorial enumerative proofs can be elegant, but they seldom are. Consider as a trivial example the problem of 1025 tennis players. Mathematically minded readers, if they haven't already heard about this famous problem, have been immediately alerted by the number. It is known to anyone who ever kept on doubling something, anything, that 1024 is 2^{10}, the product of ten factors all equal to 2. All cognoscenti know, therefore, that the presence in the statement of a problem of a number such as $1 + 2^{10}$ is bound to be a strong hint to its solution; the chances are, and this can be guessed even before the statement of the problem is complete, that the solution will depend on doubling — or halving — something ten times. The more knowledgeable cognoscenti will also admit the possibility that the number is not a hint but a trap. Imagine then that the tennis players are about to conduct a gigantic tournament in the following manner. They draw lots to pair off as far as they can, the odd man sits out the first round, and the paired players play their matches. In the second round, only the winners of the first round participate, and the whilom odd man. The procedure is the same for the second round as for the first — pair off and play at random, with the new odd man (if any) waiting it out. The rules demand that this procedure be continued, over and over again, until the champion of the society is selected. The champion, in this sense, didn't exactly beat everyone else, but he can say, of each of his fellow players, that he beat someone who beat someone, . . ., who beat someone, who beat that player. The question is: how many matches were played altogether, in all the rounds of the whole tournament?

There are several ways of attacking the problem, and even the most naive one works. According to it, the first round has 512 matches (since 1025 is odd and 512 is a half of 1024), the second round has 256 (since the 512 winners in the first round, together with the odd man of that round, make 513, which is odd again, and 256 is a half of 512), etc. The "et cetera" yields, after 512 and 256, the numbers 128, 64, 32, 16, 8, 4, 2, 1, and 1 (the very last round, consisting of only one match, is the only one where there is no odd man), and all that is necessary is to add them up. That's a simple job that pencil and paper can accomplish in a few seconds; the answer (and hence the solution of the problem) is 1024.

Mathematical students might proceed a little differently. They would quickly recognize, as advertised, that the problem has to do with repeated halvings, so that the numbers to be added up are the successive powers of 2, from the ninth down to the first, — no, from the ninth down to the zeroth! — together with the last 1 caused by the obviously malicious attempt of the problem-setter to confuse the problem-solver by using 1025 instead of 1024. The students would then proudly exhibit their knowledge of the formula for the sum of a geometric progression. They would therefore know (without carrying out the addition) that the sum of 512, 256, . . . , 8, 4, 2, and 1 is 1023, and they would then add the odd 1 to get the same total of 1024.

The trouble with the students' solution is that it's much too special. If the number of tennis players had been 1000 instead of 1025, the students would be no better off than the naive layman. The students' solution works, but it is as free of inspiration as the layman's. It is shorter but it is still, in the mathematician's contemptuous word, computational.

The problem has also an inspired solution, that requires no computation, no formulas, no numbers — just pure thought. Reason like this: each match has a winner and a loser. A loser cannot participate in any later rounds; everyone in the society, except only the champion, loses exactly one match. There are, therefore, exactly as many matches as there are losers, and, consequently, the number of matches is exactly one less than the membership of the society. If the number of members is 1025, the answer is 1024. If the number had been 1000, the answer would be 999, and, obviously, the present pure thought method gives the answer, with no computation, for every possible number of players.

Pure thought is always better than thoughtless counting. The solution of the problem by pure thought might seem less constructive than the one by counting, but it has the advantage that it can be generalized, and, for instance, it shows that the advantage gained by treating a near neighbor of a power of 2 is illusory.

The tennis player problem is a microcosmic example of an abstract and pretty piece of mathematics. The example is bad because, although mathematicians are usually interested in things other than counting, it deals with counting; it's bad because it does not, cannot, exhibit any of the conceptual power and intellectual technique of non-trivial mathematics; and it's bad because it illustrates applied mathematics (that is mathematics as applied to a "real life" problem) more than it illustrates pure mathematics (that is, the distilled form of a question about the logical interrelations of concepts — concepts, not tennis players, and tournaments, and matches). For an example, for a parable, it does pretty well nevertheless; anybody whose imagination is good enough mentally to reconstruct the ocean from a drop of water can reconstruct abstract mathematics from the problem of the tennis players.

2. A small alphabet

The so-called logistic method is the same as the combinatorial method; it is the study of a formal language by a combinatorial analysis of its letters, words, and sentences. It is the most common approach to most parts of logic; what follows is an artificial illustration of how it works.

A formal language is like a sentence machine, a computing machine, an electronic typewriter whose output is a set of sentences. The emphasis in the following description of the design of such a typewriter will not be on the wiring and the transistors that make it up, but on the list of specifications it must satisfy. The specifications are so simple that, in these days of electronic miracles, even a layman will regard it as obvious that any competent technician could construct a machine satisfying them, and that is indeed so.

The specifications for the electronic typewriter to be considered are of four major kinds: letters, words, sentences, and truths. (These are not standard technical terms — they are just a first attempt to warn the reader to expect four different levels of complication in what's coming.) Words consist of letters, sentences consist of words, and truths are special sentences; typically the definitions of word, sentence, and truth are given genetically in terms of (i) basic data, and (ii) instructions for putting old data together in new ways. When ultimately the machine becomes operational, there will be a button marked GO; when that button is pushed, the machine will proceed to print the truths, all the truths, and nothing but the truths in some prearranged (say alphabetical) order.

The alphabet of the electronic typewriter under construction consists of the four letters

D E A N.

Let these be arranged in a fixed order (the alphabetical order will do fine). A typewriter equipped with this alphabet and with the instruction to go to work is not, in the absence of any further structure, going to produce anything of interest. If the GO button is pushed, the machine can do nothing more imaginative than run through the entire (infinite) conceivable dictionary involving the alphabet. The way it might proceed is this: print (in alphabetical order) all one-letter strings, all two-letter strings, all three-letter strings, etc. Thus, on the word GO the machine will print (on successive spots of a rolling tape that is being fed through it)

A D E N AA AD AE AN DA DD
DE DN ⋯ AAA AAD AAE AAN ADA ADD ⋯.

The satisfyingly exhaustive character of this output is exceeded only by its dullness.

3. A small grammar

What makes the output just described dull is the machine's lack of selectivity; the machine prints not only the words and sentences of the language to be described, but all strings of letters, which includes the words and sentences of all conceivable languages using the same alphabet. The vocabulary of the machine — the words that it recognizes as legitimate objects of consideration — needs to be delimited.

The language that the desired machine is being designed to speak is infinitely rich in vocabulary. It has certain basic words (roots, stems) out of which all others are built, and it has certain rules of word formation (inflections) that produce new words out of old.

The basic words are the three one-letter words

E D N.

The only permissible rules of word formation are these: if a word has in it the letter D or the letter N, the string of letters obtained from that word by following it with AD or AN is also a word. This genetic definition of "word" provides a simple effective test of whether or not a string of letters is a word: if the string is not just E, peel off AD's and AN's from the right as far as possible and see whether or not the result is D or N. Example:

DANADADAN

is a word, but

DANDADAN and ANADADAN

are not.

The machine that has been taught what words are, can now be trained to do something more interesting than before, but still pretty dull: it could print the dictionary of the language. If it proceeds as before, the result will begin thus:

D E N DAD DAN NAD NAN DADAD
DADAN DANAD DANAN NADAD NADAN $\cdots$.

Words do not make a language; languages put words together to form sentences. The grammar of the language that the machine is to speak is easy to describe genetically, as follows: a sentence is obtained by writing down any word except E, following it by E, and following that by any word except E; and that is the only way to get sentences. If the typewriter is taught this definition and is then instructed to print, one after another, all sentences in increasing order of length, and, within each length, in alphabetical order, then it will begin thus:

DED DEN NED NEN DADED DADEN DANED
DANEN DEDAD $\cdots$.

4. A small logic

Grammarians might agree to stop the list of specifications for a typewriter here; all they are interested in is what a language can say and how it can say it. Logicians might want to continue, in order to learn what is truth. They are not satisfied to be able to say everything; they want to know, within each discipline, within each language, what are the true things one can say. The concept of truth in mathematics is elusive. Mathematicians usually proceed axiomatically to prove theorems. They first specify the basic data, certain sentences that are taken as true from the beginning (axioms), and then specify the instructions (rules of procedure) that produce all further provable sentences from them. The basic data of the present language consists of two sentences:

DED and NEN.

The instructions are easy too. Given a theorem — that is, a provable sentence — another theorem is obtained when any particular D that occurs in it is replaced by either

DAN or NAD,

or when any particular N that occurs in it is replaced by either

DAD or NAN.

These ways (and iterations of them) are the only ways to lengthen theorems; the only way to shorten them is to reverse the process. For example: if DANED is already known to be provable, then replace DAN by D to recapture the shorter theorem DED.

That is all there is to the list of instructions. If now the machine is told to print, successively, all theorems in order of increasing length (and alphabetically), then its production would begin with

DED	NEN	DADEN	DANED	DEDAN	DENAD
NADED	NANEN	NEDAD	NENAN	DADADED	. . . .

The construction is over. A typewriter such as the one just designed, or, better, the set of specifications for such a typewriter, complete with the description of the letters, words, sentences, and theorems, is a typical example of a logical system, of a formal logic. This is the sort of thing that logicians study all the time.

5. What is truth?

What does a logician want to know? What is a typical example of a question about formal logic? It is *not* something like this: is DADEDANADADAN provable? That is a mathematical question. The logician, the machine theorist, is interested in properties of the machine, such as whether it is large enough but not too large.

Is the machine *complete* in the sense that its output is "large enough" to include all sentences of a certain desirable form? (For example, an arbitrarily long alternating string of D's and A's that begins and ends with D always becomes a sentence when either ED or EN is tacked on to it. Is at least one of those sentences always provable?)

Is the machine *consistent*, in the sense that its output is "small enough" to exclude all sentences of a certain reprehensible form? (For instance, can both the things mentioned in the preceding example happen at the same time? That is, does it ever happen that a string like

$$\text{DADA} \cdots \text{DAD}$$

becomes a provable sentence when ED is tacked on to it, and also becomes a provable sentence when EN is tacked on to it?)

If the structure of logical systems is approached via the combinatorial method, then the techniques of logic that are employed involve close attention to symbols, their number, their order, their repeated occurrences, and their arrangements. Many parts of logic can be studied by other, algebraic, structural methods, and many concepts and relations become clearer when that is done. It must be said, however, that whichever approach is used, the study of logic does not help anyone to think, neither in the sense that it provides a source of new thoughts, nor in the sense that it enables one to avoid errors in combining old thoughts. Among the reasons for studying logic are these: (i) it is interesting in its own right, (ii) it clarifies and catalogues the methodology of mathematics, (iii) it has mathematical applications (for instance, in algebra, analysis, geometry, and set theory), and (iv) it is itself a rich application of some of the ideas of abstract algebra.

It is almost never true (but it is very pleasant when true) that the words, sentences, and theorems of a language can be defined structurally (as opposed to genetically). For the simple sample language defined above it is true. It is easy to see, for instance, that a string of letters is a word if and only if either it is the one-letter string E or it is an alternating string of consonants and A's that begins and ends with a consonant. The structural description of sentences is, for this simple language, very near to the genetic description: a string of letters is a sentence if and only if it contains exactly one E, and that E is both preceded and followed by a word other than E. The structural description of provable sentences is one very slight step further from the genetic definition: a sentence is provable if and only if the number of D's to the left of E has the same parity as the number of D's to the right of E.

6. Motivation of the small language

The reader has a right to ask, and has probably been asking, what is the motivation for studying the particular letters, words, and sentences of this language. (Motivation, not reason; the reason it was introduced was to illustrate the combinatorial method.) What, in other words, is this language talking about? The answer is: the additive properties of parity. Interpret D and N (the final letters of *odd* and *even*) as elements of a two-element additive group, that is as 1 and 0 with addition modulo 2, or, equivalently, as the quotient group of the additive group of integers modulo the subgroup of even integers. Interpret A (the initial letter of *addition*) as addition. Interpret E (the initial letter of *equality*) as equality. The words of the language then consist of one verb (E, for equals) and an infinite collection of additively combined parities (such as "odd + odd" or "odd + even + even" — denoted by DAD and DANAN respectively). The sentences of the language are equalities such as "odd + odd = even" (DADEN) and "even + even = odd" (NANED), which may or may not be provable. The basic axioms of the language are

"odd = odd" and "even = even".

The rules of procedure enable one to replace "odd" by either "odd + even" or "even + odd", to replace "even" by either "odd + odd" or "even + even", or to perform the inverses of these replacements. The result is that the theorems of the theory assert nothing more or less than that (i) an odd number of odds with any number of evens has an odd sum, and (ii) an even number of odds with any number of evens has an even sum. Since it is true that the sum of any finite number of odds is either odd or even, it follows that the linguistic description of these trivialities about parities is *complete*; since it is true that the sum of no finite number of odds is both odd and even, it follows that that same linguistic description is *consistent*.

One can also define structurally a notion of truth for the small theory of parity. A sentence UEV is *true* (under the interpretation in the two-element additive group described above) if and only if, when each occurrence of the symbol D is interpreted as 1 and each occurrence of N is interpreted as 0, then the words U and V denote the same number (either 0 or 1). Every provable sentence of the small theory is easily seen to be true. This fact is expressed by saying that the theory is *semantically sound*. In the context of the small theory, semantic soundness is equivalent to consistency. The converse of soundness is also true of the small theory: every true sentence is provable. This fact is expressed by saying that the small theory is *semantically complete*. Semantic completeness is a rarer property of theories than soundness. In the context of the small theory, semantic completeness is equivalent to completeness.

7. All mathematics

All extant mathematics can be subsumed under the theory of electronic typewriters such as the parity machine discussed above. It is convenient (but not necessary) to work with an alphabet considerably larger than four symbols; something like 10 or 15 letters are ample. The usual list of 10 or 15 letters is very likely to include

$$=,$$

$$0 \quad \text{and} \quad 1,$$

and

$$+ \quad \text{and} \quad \times.$$

In addition, the list almost certainly includes the set-theoretic symbol

$\in$ (often pronounced as “belongs to”),

and certainly something like the logical symbols

$\neg$ (often pronounced as “not”),

$\vee$ (often pronounced as “or”),

$\wedge$ (often pronounced as “and”),

$\forall$ (often pronounced as “for all”),

and

$$\exists \quad \text{(often pronounced as "there exists").}$$

The alphabet must also include a list of "variables", such as, say, the letter x, so that the language may say things like

$$(\exists x)\,(x + 1 = 0).$$

In principle infinitely many variables are needed (a finite number for any particular sentence, but an infinite supply for sentences of increasing complexity), but in fact one is enough, say x, together with a marker, such as the one in

$$x^{+},$$

whereby other variables can be manufactured, as in

$$x^{++}, \quad x^{+++}, \quad x^{++++}, \quad x^{+++++}, \quad \text{etc.}$$

These informal remarks, the description of the alphabet, are easy to convert, on demand, into a precise genetic description.

The definition of a word (best given in genetic terms) is more complicated than for the DEAN language, but it involves no new conceptual difficulties. Sample words:

$$0, \qquad 1 + 0, \qquad x^{+++} + x^{++} + 1.$$

The same comment (more complicated, but within reach) applies to the definition of sentences. Sample sentences:

$$(\exists x)\,(x + 1 = 0),$$

$$1 + 0 = 0 + 1,$$

$$0 \in \{0\},$$

$$\neg\,(0 \in \{0\}).$$

The same comment applies to the axioms (basic data) and rules of inference (instructions) too. Sample axiom:

$$(\exists x)\,(\neg\,(x = 0)).$$

Sample rule of inference: if a sentence is known to be provable, and if that sentence, preceded by $\neg$, followed by $\vee$, and then followed by another sentence, is also known to be provable, then that second (follower) sentence shall be regarded as provable also.

Once the details are filled in (they are easy to fill in, and they will be filled in below), the "large enough" and "small enough" questions can be asked. Sample: given any sentence, will the machine ever print both it, and it with $\neg$ in front? (If the answer is yes, the machine is called *inconsistent.*) This is the "too large" type of question. The answer to it for set theory and elementary number theory (the theory of addition and multiplication of natural numbers) is not known; for various fragments (such as elementary real number theory and the elementary theory of addition of natural numbers) it is known to be yes.

What about proof? What is a proof? The answer is easy: a proof is a finite sequence of sentences printed on the machine tape such that each one is either in the basic data or derivable from earlier sentences in the sequence by the instructions.

What then about "large enough" questions? Here is one: given any sentence, will the machine always print either it or it with $\neg$ in front? Equivalently: is the theory complete? Here the answer is known, and it is the subject of the celebrated Gödel incompleteness theorem. The answer is no: it is no for set theory, and it is even no for the elementary theory of natural numbers. For more restricted theories such as elementary real number theory the answer is yes.

Propositional calculus

8. Propositional symbols

An easy and basic fragment of mathematical logic is known as the *propositional calculus* (or sentential calculus). In its purely formal aspect the propositional calculus is the study of certain strings (finite sequences) of letters, made up of an alphabet (prescribed set) of six letters. Two of these "letters" are

$$\neg$$

and

$$\vee,$$

and it is probably harmless to pronounce them as "not" and "or" respectively. (The intuitive content of "or" here is the inclusive one in "Newton was brilliant or lucky", not the exclusive one in "Newton was handsome or ugly".) The next two letters are

$$p$$

and

$$+,$$

where the latter is just a suffix suitable for repeated attaching to p, as in p^{++++}. The letters $\neg$ and $\vee$ are *propositional connectives*.

Only one kind of admissible formula needs to be considered (not both "words" and "sentences"). The "sentences" of the propositional calculus are obtained as follows. (i) The letter p is a sentence, and it remains so when followed by an arbitrary finite number of occurrences of the symbol "+"; the sentences obtained in this way are called *propositional variables*. (ii) If x is a sentence, the result of placing $\neg$ immediately to the left of x is also one. (iii) If x and y are sentences, the result of following x by $\vee$ and that by y is also a sentence. All sentences are obtained from propositional variables by repeated applications of the latter two methods. More or less equivalently, the set of all variables is the smallest set of finite sequences that contains the one-termed sequence p and that is closed under the operation of suffixing $^{+}$; the set of all sentences is the smallest set of finite sequences that contains all variables and is closed under the operations of prefixing $\neg$ and infixing $\vee$. Incidentally, sentences are frequently called *well-formed formulas*, abbreviated wff.

The preceding paragraph is wrong. What is wrong is illustrated by the following string:

$$\neg p \vee p.$$

Is it a sentence? Since p is a sentence (by (i)), and since $\neg p$ is a sentence (by (ii)), the answer is yes (by (iii)). Alternatively, since p and p are sentences (by (i)), and since therefore $p \vee p$ is a sentence (by (iii)), the answer is yes (by (ii)). These two constructions of $\neg p \vee p$ correspond to intuitively distinct interpretations. If p denotes "Newton was brilliant", then the result of the first construction says that Newton was either brilliant or not, whereas the result of the second construction says that he was not. The chief way a definition can be wrong is by failing to define what the definer intended; that is how the preceding paragraph is wrong. If parentheses are used in the normal mathematical way, then the distinction between the two constructions leading to $\neg p \vee p$ is that between

$$(\neg p) \vee p \quad \text{and} \quad \neg(p \vee p).$$

This is clear — but it uses symbols not in the list of four given so far. A remedy is also clear: give official recognition to the clarifying symbols. The propositional calculus, as it is to be studied in the sequel, has six letters; the missing two are

$$($$

and

$$).$$

The proper way to define "sentence" is to replace (ii) by: if x is a sentence, so is $(\neg x)$, and to replace (iii) by: if x and y are sentences, so is $(x \vee y)$. These replacements (as well as the original (i)–(iii)) are actually abbreviations. The unabbreviated version of the new (ii), for instance, reads as follows: if x is a sentence, then so is the sequence whose first two terms, in order, are

$$($$

and

$$\neg,$$

whose following terms are those of x, in order, and whose last term, immediately following the last term of x, is

$$).$$

The sentence $(\neg x)$ is called the *negation* of the sentence x, and the sentence $(x \vee y)$ is called the *disjunction* of the sentences x and y. Here are some sample sentences of the propositional calculus:

$$((\neg(p \vee p)) \vee p),$$

$$((\neg p) \vee (p \vee p^{+})),$$

$$((\neg(p \vee p^{+})) \vee (p^{+} \vee p)),$$

$$((\neg((\neg p) \vee p^{+})) \vee ((\neg(p^{++} \vee p)) \vee (p^{++} \vee p^{+}))),$$

$$((\neg p) \vee (p \vee (\neg p))),$$

$$((\neg((\neg(p \vee p^{+})) \vee p^{++})) \vee p^{++}).$$

9. Propositional abbreviations

It is often typographically and psychologically convenient to abbreviate some of the strings of symbols that occur in the propositional calculus, that is, to replace them by other (shorter) strings. The simplest abbreviations are alphabetical: use some other letter of the alphabet, for example, q and r in place of p^{+} and p^{++} respectively. This will be done systematically in what follows. (Other alphabetic substitutions are occasionally convenient, and might be used without any further apology.)

As a somewhat different kind of abbreviation, we introduce a new connective. It is denoted by the symbol

$$\Rightarrow$$

and is defined, whenever x and y are sentences, by saying that

$$(x \Rightarrow y)$$

means

$$((\neg x) \vee y).$$

The symbol $\Rightarrow$ may be pronounced "implies" (x implies y) or "if ... then ..." (if x, then y), and the sentence $(x \Rightarrow y)$ is called an *implication*. The intuitive background of the replacement is illustrated by the relation between the meanings of the ordinary English sentences "if this coin comes up heads, we owe you a dollar" and "either the coin does not come up heads, or we owe you a dollar". An alternative motivation for the replacement is in thinking about the negation of an implication and comparing it with the negation of the corresponding disjunction.

A third and purely typographical kind of abbreviation concerns the omission of parentheses from a sentence. For example, the outermost pair of parentheses will be omitted from a sentence when it is discussed in isolation from all others. Parentheses were introduced to make unambiguous the growth of a sentence, and are unnecessary once a sentence is fully grown. In arithmetic there are conventions regarding the precedence of operations that allow certain parentheses to be omitted. Thus, the expression $((-x) \cdot y) + z$ may be unambiguously written as $-x \cdot y + z$. There are similar conventions regarding the precedence of operations in logic: negation has precedence over all binary operations, and, among binary operations, disjunction has precedence over implication.

The sample sentences given above, rewritten in the abbreviated notation proposed in the preceding three paragraphs, take the form:

$$p \vee p \Rightarrow p,$$

$$p \Rightarrow p \vee q,$$

$$p \vee q \Rightarrow q \vee p,$$

$$(p \Rightarrow q) \Rightarrow (r \vee p \Rightarrow r \vee q),$$

$$p \Rightarrow p \vee \neg p,$$

$$(p \vee q \Rightarrow r) \Rightarrow r.$$

10. Polish notation

There is an altogether different version of the propositional calculus, the one expressed in the parenthesis-free notation (frequently called the Polish notation) of Łukasiewicz. (Just what it means to say that what follows is another version of what has preceded has not been explained so far; the most convenient place for the explanation is after the full development of the propositional calculus.) One way to describe the Polish notation is to return to the four symbols $\neg$, $\vee$, p, and $^{+}$ — that is, temporarily to forget about the afterthought parentheses. (It is, however, still pleasant to use $p, q, \ldots$ instead of $p, p^{+}, \ldots$, and that sort of abbreviation will be retained.) The new definition of sentence replaces the old (corrected) (iii) by: if x and y are sentences, then so is

$$\vee xy.$$

(More precisely: if x and y are sentences, then so is the sequence whose first term is $\vee$, whose following terms are those of x, in order, and whose following terms after that are those of y, in order. The old, unabbreviated (i) and (ii) remain in force, unaltered.) What was $((\neg p) \vee p)$ becomes

$$\vee \neg pp,$$

whereas $(\neg (p \vee p))$ becomes

$$\neg \vee pp.$$

Generally, it could be said that Polish sentences are written from right to left, or from outside in. More to the point is that they are deciphered (transcribed to the parenthesis notation) from inside out. That is: reading from right to left, find the first connective; if it is $\neg$, and if the propositional variable to the right of it is, say, s, then begin the transcription by $(\neg s)$, but if it is $\vee$, and if the two propositional variables immediately to the right of it are, say, r and t, then begin the transcription by

$$(r \vee t).$$

Continue the same way, reading from right to left, and treating already transcribed fragments the same way as isolated propositional variables. The six sample sentences used before now become:

$$\vee \neg \vee ppp,$$

$$\vee \neg p \vee pq,$$

$$\vee \neg \vee pq \vee qp,$$

$$\vee \neg \vee \neg pq \vee \neg \vee rp \vee rq,$$

$$\vee \neg p \vee p \neg p,$$

$$\vee \neg \vee \neg \vee pqrr.$$

Actually, Łukasiewicz used symbols different from $\vee$ and $\neg$ for the propositional connectives, but this minor notational difference is of no importance.

The Polish notation has some technical advantages, but, for most people, intuitive ease of reading is not one of them. It is worth remarking though that it has wider applications than indicated above. Example: the parenthesized distributive law

$$((a \cdot (b + c)) = ((a \cdot b) + (a \cdot c)))$$

can also be written as the parenthesis-free

$$= \cdot a + bc + \cdot ab \cdot ac.$$

11. Language as an algebra

Negation and disjunction are, in effect, operations on the set of sentences of the propositional calculus. That is, negation is a unary operation that associates with every sentence x the sentence $\neg x$, and disjunction is a binary operation that associates with every two sentences x and y the sentence $x \vee y$. In a slight abuse of notation, the symbols $\neg$ and $\vee$ are used to denote, not only the letters "not" and "or", but also the operations of negation and disjunction. Thus, there is associated with the propositional calculus a natural algebraic structure whose universe is the set of all sentences and whose operations are negation and disjunction.

12. Concatenation

Much of logical pedantry has to do with the concept of concatenation. If x and y are finite sequences (of letters of the alphabet, or of anything), then purists are bothered by symbols such as xy; it is, to them, not at all obvious that such a symbol must denote the sequence that begins as x does and continues with the terms of y. (That the preceding sentence commits the error of using a thing as a name of that thing is just adding wanton insult to cruel injury.) The objection is valid; nothing means something without a definition. The linguistic precision and virtuosity that logicians are forced to employ is caused by their completely justified insistence on clarity; the concept of sequential concatenation, though perhaps an easy one, is complicated to describe in mere words. What is sometimes overlooked, however, is that there is available to mathematicians a language other than erroneous symbolism and pedantic circumlocution; the name of that languge is mathematics. Perhaps some authors are frightened by the threat of circularity; logic, they might think, is here to serve mathematics, and it is inadmissible and dangerous to use mathematics in the description of logic. There is much to be said against that view, but to say it would be a digression here. Suffice it to say that, from the point of view of this exposition, logic is not something to base mathematics on, but a branch of mathematics, and the application of normal mathematical language is perfectly fair.

What then is a sentence in the propositional calculus, or in any other logistic system? It is a certain kind of finite sequence (of letters and other symbols) — and what is a finite sequence? The answer is clear in mathematical terms: it is a function x whose domain is the set of natural numbers smaller than some fixed natural number n, that is, the set $\{0, \dots, n-1\}$. It is customary to denote the value of x at i by x_i, where $0 \leq i < n$. It is useful to allow the possibility $n = 0$ also; in that case x is the empty function with, naturally, the empty domain and, correspondingly, the empty range.

There is a natural binary operation on finite sequences, an ad hoc (and temporary) symbol for which might be $\frown$, defined as follows. If x and y are finite sequences, so is

$$x^\frown y;$$

in more detail, if x and y are of lengths m and n respectively (that is, with domains consisting of the predecessors of m and n), then $x^\frown y$ is of length $m + n$. The values (terms) of $x^\frown y$ are given by

$$(x^\frown y)_i = x_i \quad \text{if} \quad i < m,$$

and

$$(x^\frown y)_{m+i} = y_i \quad \text{if} \quad i < n.$$

The name of this operation is *concatenation*. Once this operation is precisely defined and well understood, there is no harm in simplifying the notation used to denote it. There is, in particular, no harm in denoting $x^\frown y$ by

$$xy,$$

that is, in using concatenation (the English word) to denote concatenation (the technical term).

The operation of concatenation, acting on the set of all finite sequences of, say, letters from a finite alphabet A (that is, elements from a fixed finite set A) is associative. (If x, y, and z are finite sequences, then

$$(xy)z = x(yz);$$

the parentheses in this equation are not parts of the alphabet but directions describing the order in which the concatenations are to be formed.) The set of all non-empty finite sequences of A is called the *free semigroup* generated by A. ("Semigroup" refers to an associative binary operation; "free" refers to a technical algebraic property, to be defined later, of this particular semigroup. Other manifestations of that property will occur and will be discussed below.) If the empty sequence is re-admitted, the semigroup acquires a *unit*; this means that if $\varnothing$ is the empty sequence, and if x is an arbitrary finite sequence, then both $\varnothing x$ and $x\varnothing$ are equal to x.

13. Theorem schemata

Some of the sentences of the propositional calculus are singled out and called *theorems*. The usual way of describing the set of all theorems as a subset of the set of all sentences is via "axioms" (or axiom schemata), and "rules of procedure" (or rules of inference). An axiom, in this sense of the word, is a particular sentence of the propositional calculus; an axiom schema is a set of sentences, usually described as the set of all sentences of a certain form. ("Schema" is the singular; "schemata" is the plural.) A rule of procedure is an operation (unary, binary, ...) on sentences that makes new sentences out of old. The axiom-rule way of specifying theorems is to exhibit certain sentences and to dub them theorems by fiat, and then to define the set of all theorems as the smallest set of sentences that contains the axioms and is closed under the action of the rules.

A frequently used set of axioms consists of the first four of the six sentences that were repeatedly used as examples above. Here they are, again:

$$p \vee p \Rightarrow p,$$

$$p \Rightarrow p \vee q,$$

$$p \vee q \Rightarrow q \vee p,$$

$$(p \Rightarrow q) \Rightarrow (r \vee p \Rightarrow r \vee q).$$

Recall now that the propositional calculus has an infinite supply of propositional variables, namely p, $q\ (= p^{+})$, $r\ (= p^{++})$, $s\ (= p^{+++})$, etc. When $p \vee p \Rightarrow p$ is offered as an axiom, an inexperienced reader in a hurry might think of p here as a "variable" (whatever that may be), and might consequently think that once "$p \vee p \Rightarrow p$" is included among the axioms, "$q \vee q \Rightarrow q$" (and all others of the same form) come automatically along with it. This is not so, however, because the variables do not really vary. The only way to achieve this desideratum (and it *is* the desideratum) is explicitly to say it. There are two natural ways of saying it. One is via the axiom schema $x \vee x \Rightarrow x$. What is intended here is more precisely expressed as follows: for all sentences x, the sentence $x \vee x \Rightarrow x$ (that is, the concatenation of x, $\vee$, x, $\Rightarrow$, and x, in that order — or, better, the concatenation of

$$(, (, \neg, (, x, \vee, x,),), \vee, x,)$$

in that order) is, by definition, a theorem. Another way of achieving the same result is to leave the single axiom as is, and to consider a substitution rule. The operation of substitution is defined as follows: if x and y are sentences and u is a variable, then $x[u/y]$ is the result obtained from x by replacing each occurrence of u by y. (A convenient way of reading the abbreviation is: "substitute y for u in x".) Example:

$$(p \vee p \Rightarrow p)[p/x] = (x \vee x \Rightarrow x).$$

Here is a slightly more involved example:

$$(\neg (p \vee p) \vee p)[p/(p \vee q)] = (\neg ((p \vee q) \vee (p \vee q)) \vee (p \vee q)).$$

To recapture the same set of theorems as is described by the schema

$$x \vee x \Rightarrow x$$

all that is necessary is to say that, by definition,

$$(p \vee p \Rightarrow p)[p/x]$$

is a theorem whenever x is a sentence. The difference between the two methods is not very big; one describes a set as all elements of a certain form (schema), and the other describes the same set as the range of a certain function (rule).

Similar remarks apply, of course, to the other axioms. Thus, for instance, the second axiom can be replaced by the schema $x \Rightarrow x \vee y$. More precisely: for all sentences x and y, the sentence $x \Rightarrow x \vee y$ is, by definition, an axiom. These axioms could be described by a substitution rule similar to the one introduced above, but the situation is combinatorially more complicated than it might first appear. The literature is full of mistakes in the formulation of substitution rules. The sort of thing that goes wrong is illustrated by the sentence

$$p \vee q \Rightarrow (p \vee q) \vee (r \vee q).$$

This is the result of letting x be $p \vee q$ and y be $r \vee q$ in the schema $x \Rightarrow x \vee y$. Alternatively, one might try to replace p in

$$p \Rightarrow p \vee q$$

by $p \vee q$ and then to replace q by $r \vee q$ in the result. The outcome would not be what is intended; it would be

$$p \vee (r \vee q) \Rightarrow ((p \vee (r \vee q)) \vee (r \vee q)).$$

(To make the confusion minimal, several parentheses were added along with the indicated strings of letters.) The way out of the difficulty is to reach out for auxiliary, irrelevant, propositional variables along the way, and make the necessary substitutions one at a time, avoiding all danger of collision. To say this clearly and precisely is more trouble than it is worth. By far the simplest thing to do is to present the axioms of the propositional calculus as schemata, and, in the sequel, that is how the propositional calculus will be regarded. Thus, if x, y, and z are sentences, then each of the sentences

T1 $$x \vee x \Rightarrow x,$$

T2 $$x \Rightarrow x \vee y,$$

T3 $$x \vee y \Rightarrow y \vee x,$$

T4 $$(x \Rightarrow y) \Rightarrow (z \vee x \Rightarrow z \vee y)$$

is an axiom of the propositional calculus, and all axioms are one of these. (The letter "T" at the left of these schemata is an abbreviation of "theorem".)

The definition of the set of theorems for the propositional calculus is not yet complete; one rule (but not a substitution rule) is necessary. That rule, μ, a binary one, is known as *modus ponens*, and can be described as follows: if x and z are sentences and if z has the form $x \Rightarrow y$ (that is, $\neg x \vee y$), then $\mu(x, z) = y$; otherwise $\mu(x, z) = x$. The latter clause ("otherwise $\mu(x, z) = x$") is, from the practical point of view, unnecessary and is usually not stated; it was added here just to make μ an everywhere defined binary operation on sentences. There are various ways of avoiding it. One way is simply to resign one's self to binary operations that are not defined for all possible pairs of arguments; another is to regard a rule of inference not as a function but as a relation. The latter point of view is probably the healthiest. From that point of view, modus ponens is the assertion that *whenever x and y are sentences such that x and $x \Rightarrow y$ are theorems, then y is a theorem.*

A complete definition of theorem can now be formulated as follows. A sentence y is a theorem if and only if either

(i) y is an axiom,

or

(ii) there exist theorems x and z such that $z = (x \Rightarrow y)$.

In more detail (and with less seeming circularity): y is a theorem if and only if there exists a finite sequence of sentences, whose last term is y, such that each sentence in the sequence is either (i) an axiom, or (ii) the result of applying modus ponens to some pair of preceding sentences in the sequence. A finite sequence with these properties is called a *proof* (a formal proof, a formal proof of y).

14. Formal proofs

The construction and the study of formal proofs is usually more boring than profitable and more time-consuming than instructive. To prove this statement, all that is necessary is to construct and to study a few formal proofs. Here are two samples.

Formal proof I.

$$(p \Rightarrow q) \Rightarrow (\neg r \vee p \Rightarrow \neg r \vee q).$$

(Use T4, with $x = p$, $y = q$, $z = \neg r$.) Even this trivial, one-line, formal proof helps to show that formal proofs might leave something to be desired. The point is that the method is much more general than the result it was used to produce. The one line is an instance of T4. If x, y, and z are arbitrary sentences, then

T5 $$(x \Rightarrow y) \Rightarrow ((z \Rightarrow x) \Rightarrow (z \Rightarrow y))$$

is always an instance of T4 (just replace z by $\neg z$ and recall that $z \Rightarrow x$ is an abbreviation for $\neg z \vee x$). This kind of statement about the propositional calculus may be called a *derived axiom schema*; in what follows we shall usually feel just as free to use such derived schemata as if they had been postulated in the beginning. Whenever that is done, however, the context loses the right to be called a formal proof. In formal proofs no provision exists for using the result of an earlier formal proof; the only permissible anterior constituents are the axioms and the results of applying modus ponens to anterior constituents.

Formal Proof II.

(1) $p \Rightarrow p \vee p$ (T2 with $x = p$, $y = p$)

(2) $p \vee p \Rightarrow p$ (T1 with $x = p$)

(3) $(p \vee p \Rightarrow p) \Rightarrow ((p \Rightarrow p \vee p) \Rightarrow (p \Rightarrow p))$

(T4 with $x = p \vee p, y = p$, $z = \neg p$)

(4) $(p \Rightarrow p \vee p) \Rightarrow (p \Rightarrow p)$ (Modus ponens, (2) and (3))

(5) $p \Rightarrow p$ (Modus ponens, (1) and (4)).

(The numbers on the left and the parenthetical explanations on the right are not parts of the formal proof; the former are for use in the latter, and the latter are to save the reader the time it takes to discover the justification for including in the formal proof each particular step.) Note that the conclusion (line (5)) of this formal proof can (in fact should) be written in the form

$$\neg p \vee p.$$

In the proof there occurs a pattern that occurs in many formal proofs. The first two steps in the proof have the form $x \Rightarrow y$ and $y \Rightarrow z$. Since

$$(y \Rightarrow z) \Rightarrow ((x \Rightarrow y) \Rightarrow (x \Rightarrow z))$$

(or, rather, any instance of it) is an instance of T4, it is fit to appear in any formal proof. From it, and from $x \Rightarrow y$ and $y \Rightarrow z$, two applications of modus ponens yield $x \Rightarrow z$.

These remarks show that *whenever x, y, and z are sentences such that $x \Rightarrow y$ and $y \Rightarrow z$ are theorems, then $x \Rightarrow z$ is a theorem.* This kind of statement about the propositional calculus is sometimes called a *derived rule of inference* (or metatheorem). In ordinary mathematical language it is, of course, a theorem about the propositional calculus; the theorem may be loosely described (and accurately referred to) as the *transitivity of implication*. The use of this theorem and of sentences of the form

$$x \Rightarrow x \vee x$$

(T2 with $y = x$) and

$$x \vee x \Rightarrow x$$

(T1) yields the result that

T6 $$x \Rightarrow x$$

is a theorem, whatever the sentence x may be; in other words, T6 is another derived axiom schema.

Observe that T6 is an abbreviation; in less abbreviated form, the set of theorems it describes consists of all sentences of the form $\neg x \vee x$. It is natural to guess (on both intuitive and formal grounds) that

T7 $$x \vee \neg x$$

is also a theorem for all x, and that is indeed so. Reason: apply modus ponens to T6 and T3 (in the latter replacing x and y by $\neg x$ and x respectively). This argument (the preceding sentence) is mathematically perfectly respectable, but it is not a formal proof in the sense in which that phrase was defined. It cannot be: a formal proof consists of sentences, in particular its last term is a sentence, and $x \vee \neg x$ is not a sentence; it uses letters that are not part of the formal alphabet. Here is how a closely related formal proof would look.

$$p \Rightarrow p \vee p,$$

$$p \vee p \Rightarrow p,$$

$$(p \vee p \Rightarrow p) \Rightarrow ((p \Rightarrow p \vee p) \Rightarrow (p \Rightarrow p)),$$

$$(p \Rightarrow p \vee p) \Rightarrow (p \Rightarrow p),$$

$$p \Rightarrow p,$$

$$(p \Rightarrow p) \Rightarrow (p \vee \neg p),$$

$$p \vee \neg p.$$

Explanation: the first five lines of the preceding seven are a word-for-word copy of formal proof II; the next line is an instance of T3 (with $x = \neg p$ and $y = p$), and the last line is the result, via modus ponens, of its two predecessors. Since the process of copying formal proofs from one place to another, and the related process of repeating step-by-step special cases of a procedure that is known in much greater generality (in the form of rules of inference) are manifestly time-consuming and unprofitable, and since, at best, these repetitious processes yield a particular sentence (as opposed to an infinite class of sentences of a certain form), we hereby abandon formal proofs in the propositional calculus forever. What follows is a rigorous (but not, in the technical sense, formal) derivation of some further sets of theorems and some further rules of inference.

It will be convenient to have a quick way of expressing the statement that a certain sentence x is a theorem; a standard shorthand is

$$\vdash x.$$

Accordingly, the axiom schemata T1–T4 and the derived schemata T5–T7 could and should have been preceded by the symbol $\vdash$.

The next five derived theorem schemata to be considered are

T8 $\qquad \vdash x \Rightarrow x \vee x,$

T9 $\qquad \vdash x \Rightarrow y \vee x,$

T10 $\qquad \vdash x \Rightarrow \neg\neg x,$

T11 $\qquad \vdash \neg\neg x \Rightarrow x,$

T12 $\qquad \vdash (x \Rightarrow y) \Rightarrow (\neg y \Rightarrow \neg x).$

The proofs are straightforward. For T8: apply T2, with x in place of y. For T9: apply T2, T3, and the transitivity of implication. For T10: apply T7 to $\neg x$ in place of x. For T11:

$$\vdash \neg x \Rightarrow \neg\neg\neg x$$

by T10, whence

$$\vdash x \vee \neg x \Rightarrow x \vee \neg \neg \neg x$$

by T4; in view of T7 and modus ponens, it follows that

$$\vdash x \vee \neg \neg \neg x$$

and hence, by T3 and modus ponens, that

$$\vdash \neg \neg \neg x \vee x.$$

The latter is T11 (unabbreviated). For T12:

$$\vdash y \Rightarrow \neg \neg y$$

by T10, whence

$$\vdash \neg x \vee y \Rightarrow \neg x \vee \neg \neg y$$

by T4 and modus ponens;

$$\vdash \neg x \vee \neg \neg y \Rightarrow \neg \neg y \vee \neg x$$

by T3;

$$\vdash \neg x \vee y \Rightarrow \neg \neg y \vee \neg x$$

by the transitivity of implication. The latter is T12 (somewhat unabbreviated).

Many of the theorems here proved have names. For instance, T1 and T8 are called the *idempotence laws for disjunction*, T2 and T9 the *absorption laws for disjunction*, T3 the *commutative law for disjunction*, T4 the *monotony law for disjunction*, T6 and T7 the *laws of the excluded middle*, T10 and T11 the *laws of double negation*, and T12 the *law of contraposition*.

15. Entailment

The time has come to return to the algebraic point of view. As was mentioned before, the set of sentences of the propositional logic, under the operations of negation and disjunction, forms an algebra. Implication is a derived (binary) operation of this algebra in the sense that it is defined in terms of the primitive operations, in this case negation and disjunction.

It is now convenient to introduce a binary relation, to be called *entailment*. The relation (symbolized by $\leq$) is defined for sentences; if x and y are sentences, then

$$x \leq y$$

means, by definition, that $x \Rightarrow y$ is a theorem (that is, that $\vdash x \Rightarrow y$). In the study of entailment it is more important than ever to resist the temptation to make the propositional calculus talk. The "sentences" of the propositional calculus are elements of an algebraic structure (from the present point of view even the fact that they are strings of letters is irrelevant), and the resemblance of the symbolism to various more or less standard abbreviations of common sense logic (for example, $x \Rightarrow y$ for "x implies y") is a coincidence. That the coincidence is intentional does not lessen the danger of (and the need to avoid) confusion. An expression such as $x \leq y$ (where x and y are sentences) does talk; it is an assertion in ordinary mathematese (namely the assertion that the sentence $x \Rightarrow y$ belongs to a certain unambiguously if lengthily defined set of sentences called theorems). For a close and suggestive analogy, suppose that within the set of real numbers the strict concepts of positiveness and negativeness and the corresponding weak concepts of non-negativeness and non-positiveness are already known, but that the usual ordering of real numbers has not yet been defined. In that case it would be quite natural to define the binary relation $x < y$ (for numbers) to mean that $x - y$ (the numerical difference) is negative. (Similarly, $x \leq y$ could be defined to mean that $x - y$ is non-positive.) A very slightly more complicated, and correspondingly slightly deeper and more usable, analogy concerns vectors in the real plane. It is feasible and useful to define an order relation for such vectors by saying that $x \leq y$ means that $x - y$ is in, say, the left half plane (boundary included). In these analogies the order relation ($\leq$) corresponds to entailment ($\leq$), the algebraic operation of subtraction ($-$) corresponds to implication ($\Rightarrow$), and the prescribed set of numbers and vectors (the non-positive ones) correspond to the prescribed set of sentences (the theorems).

The symbol for entailment resembles symbols for order relations, and the analogies used above referred to order relations. Does entailment have the three basic properties (reflexivity, transitivity, and antisymmetry) of an order relation? The answer is yes for the first two and, interestingly, no for the third. Reflexivity says that $x \leq x$, and that is exactly the assertion of T6; transitivity says that if $x \leq y$ and $y \leq z$, then $x \leq z$, and that is exactly the assertion of the transitivity of implication. Antisymmetry would say that if $x \leq y$ and $y \leq x$, then $x = y$, and there are many examples to show that this is false. Thus, for one example, for any sentence x, both

$$x \vee x \leq x$$

(by T1) and

$$x \leq x \vee x$$

(by T8), but the sentence x is not the same as the sentence $x \vee x$. (The latter is, for one thing, more than twice as long as the former.) For another example consider

$$x \leq \neg \neg x$$

(T10) and

$$\neg \neg x \leq x$$

(T11). In the vector analogy, the simultaneous validity of $x \leq y$ and $y \leq x$, does not imply that $x = y$, but merely that x and y have the same abscissa.

16. Logical equivalence

The situation we have thus arrived at is a quite common one in mathematics; it calls for the introduction of a new binary relation. The new relation (symbolized by $\equiv$) is called *equivalence* (or logical equivalence). If x and y are sentences, then

$$x \equiv y$$

means, by definition, that $x \leq y$ and $y \leq x$, that is, that both $x \Rightarrow y$ and $y \Rightarrow x$ are theorems. The examples of the preceding paragraph can now be expressed by writing

$$x \vee x \equiv x$$

(idempotence of disjunction) and

$$x \equiv \neg \neg x$$

(double negation). It is obvious that logical equivalence is indeed an equivalence relation (reflexive, symmetric, transitive). One of the main purposes of what follows is to show that logical equivalence is also a congruence relation on the algebra of sentences, that is, it is related in an algebraically fruitful way to the propositional connectives.

The simplest way to study the relation between equivalence and the connectives is first to study the relation between entailment and the connectives. Suppose therefore that $x \leq y$. Since this means that

$$\vdash x \Rightarrow y,$$

it follows (T12 and modus ponens) that

$$\vdash \neg y \Rightarrow \neg x.$$

In other words,

$$\text{if } x \leq y, \text{ then } \neg y \leq \neg x$$

(contraposition). Apply this twice (to $x \leq y$ and to $y \leq x$), and conclude that if $x \equiv y$, then $\neg y \equiv \neg x$; logical equivalence is compatible with negation.

If, again, $x \leq y$, so that $\vdash x \Rightarrow y$, then, whenever z is a sentence,

$$\vdash z \vee x \Rightarrow z \vee y$$

(by T4 and modus ponens), so that

$$z \vee x \leq z \vee y.$$

Apply this twice and conclude that if $x \equiv y$, then

$$z \vee x \equiv z \vee y;$$

logical equivalence is compatible with disjunction on the left. Since, by T3,

$$x \vee z \leq z \vee x$$

for all x and z, it follows that if $x \leq y$, then

$$x \vee z \leq z \vee x \leq z \vee y \leq y \vee z,$$

so that

$$x \vee z \leq y \vee z.$$

Apply this twice and conclude that logical equivalence is compatible with disjunction on the right. (Note: the proof did not show, but could have been used to show, that $x \vee z \equiv z \vee x$ for all x and z.)

The preceding paragraph implies that if x and y are logically equivalent sentences, if we form a sentence $u(x)$ by starting with x and any collection of other sentences and repeatedly combining them by negation and disjunction, and if $u(y)$ is the sentence obtained when the same steps are applied to y (and the same collection of other sentences), then $u(x) \equiv u(y)$. Otherwise said: if $u(x)$ is a sentence in which x occurs as a constituent, and if $u(y)$ is obtained from $u(x)$ by the replacement of y for x at any one of the occurrences of x, then

$$u(x) \equiv u(y).$$

It is useful to emphasize that in the latter (replacement) formulation, the replacement need not be performed simultaneously for all occurrences of x in $u(x)$. Thus, for instance, if $x \equiv y$, and if z is any other sentence, then

$$(z \vee x \Rightarrow \neg x) \equiv (z \vee y \Rightarrow \neg x),$$

and

$$(z \vee x \Rightarrow \neg x) \equiv (z \vee x \Rightarrow \neg y),$$

and

$$(z \vee x \Rightarrow \neg x) \equiv (z \vee y \Rightarrow \neg y).$$

The replacement property of logical equivalence, proved in the preceding paragraph and discussed in the present one, may be described by saying that logical equivalence is a congruence relation with respect to the propositional connectives.

17. Conjunction

It is now convenient to introduce, as a new abbreviation, another propositional connective (and, almost immediately thereafter, one of its descendants). The new connective is denoted by

$$\wedge$$

(pronounced "and"), and defined, whenever x and y are sentences, by the stipulation that

$$x \wedge y$$

means

$$\neg(\neg x \vee \neg y).$$

The sentence $x \wedge y$ is called the *conjunction* of x and y. The descendant of conjunction referred to above is *bi-implication* (also called the *biconditional*). The connective is denoted by

$$\Leftrightarrow$$

(pronounced "if and only if") and is defined by stipulating that

$$x \Leftrightarrow y$$

means

$$(x \Rightarrow y) \wedge (y \Rightarrow x).$$

Both conjunction and bi-implication are binary operations on the set of all sentences of the propositional calculus. Just as implication, they are derived operations in the sense that they are defined in terms of the basic operations of negation and disjunction. It follows that $\equiv$ is a congruence relation with respect to conjunction and bi-implication.

The following conventions make it possible to reduce the number of parentheses needed. Negation has precedence over both conjunction and bi-implication, and conjunction has precedence over all other binary connectives.

Since bi-implication has conjunction in its very definition, it is advisable to establish the properties of "and" before tackling the study of "if and only if", and we proceed to do so.

Here is a small batch of schemata of theorems involving conjunction:

T13 $$\vdash \neg(x \wedge y) \Rightarrow \neg x \vee \neg y,$$

T14 $$\vdash \neg x \vee \neg y \Rightarrow \neg(x \wedge y),$$

T15 $$\vdash \neg(x \vee y) \Rightarrow \neg x \wedge \neg y,$$

T16 $$\vdash \neg x \wedge \neg y \Rightarrow \neg(x \vee y).$$

All these assertions are known as *De Morgan's laws*. The proofs are easy. T13 and T14 together say that

$$\neg(x \wedge y) \equiv \neg x \vee \neg y.$$

Since $\neg(x \wedge y)$ is $\neg\neg(\neg x \vee \neg y)$ (by definition) the desired conclusion follows from the law of double negation. Similarly, T15 and T16 together say that

$$\neg(x \vee y) \equiv \neg x \wedge \neg y.$$

For the proof, replace x and y in the definition of conjunction by $\neg x$ and $\neg y$ respectively, and use the law of double negation.

The De Morgan laws, together with the law of double negation, yield an important duality principle for the propositional calculus. Consider two sentence schemata, u and v, in which the only propositional connectives that occur are negation, disjunction, and conjunction. Let u^* and v^* be the schemata obtained from u and v by interchanging all occurrences of $\vee$ and $\wedge$, that is, by simultaneously replacing all occurrences of $\vee$ by $\wedge$, and all occurrences of $\wedge$ by $\vee$. Then u^* and v^* are called the *duals* of u and v, the implication $v^* \Rightarrow u^*$ is the *dual* of the implication $u \Rightarrow v$, the bi-implication $u^* \Leftrightarrow v^*$ is the *dual* of the bi-implication $u \Leftrightarrow v$, and so on. Notice that the dual of the dual of a schema is the original schema. The *duality principle for implication* asserts that if each instance of $u \Rightarrow v$ is a theorem, then so is each instance of its dual. To prove this metatheorem (theorem about theorems), assume that

$$\vdash u \Rightarrow v.$$

Let $u^{\#}$ and $v^{\#}$ denote the result of replacing x with $\neg x$ for each variable x occurring in u and v. Then $u^{\#} \Rightarrow v^{\#}$ is a special case of $u \Rightarrow v$, and hence provable by assumption. Therefore,

$$\vdash \neg v^{\#} \Rightarrow \neg u^{\#},$$

by the law of contraposition. By repeated application of the De Morgan laws and the law of double negation, $\neg v^{\#}$ is logically equivalent to v^* and $\neg u^{\#}$ is logically equivalent to u^*. Thus

$$\vdash v^* \Rightarrow u^*.$$

Two corollaries of the duality principle for implication are the duality principles for entailment and logical equivalence:

$$\text{if} \quad u \leq v, \quad \text{then} \quad v^* \leq u^*,$$

and

$$\text{if} \quad u \equiv v, \quad \text{then} \quad u^* \equiv v^*.$$

As an immediate consequence of the duality principle, the commutative law for disjunction yields the commutative law for conjunction:

T17 $$\vdash (x \wedge y) \Rightarrow (y \wedge x).$$

Similarly, the absorption laws for disjunction yield the absorption laws for conjunction:

T18 $$\vdash x \wedge y \Rightarrow x,$$

T19 $$\vdash x \wedge y \Rightarrow y,$$

and the idempotence laws for disjunction yield the idempotence laws for conjunction:

T20 $$\vdash x \Rightarrow x \wedge x,$$

T21 $$\vdash x \wedge x \Rightarrow x.$$

Suppose now that x is a theorem and y is a sentence such that $x \leq y$. Since $\vdash x \Rightarrow y$, it follows (modus ponens) that y is a theorem; theorems entail theorems only. In the reverse direction, suppose that x is an arbitrary sentence and y is a theorem; since

$$\vdash y \Rightarrow (x \Rightarrow y)$$

(by T9, with y in place of x and $\neg x$ in place of y, and the definition of implication), it follows from modus ponens that

$$\vdash x \Rightarrow y,$$

and hence that

$$x \leq y.$$

The conclusion says that everything entails a theorem. From this observation it follows that all theorems are logically equivalent (since each of a pair of theorems entails the other). If x and y are theorems, then $x \equiv y$, by the previous remark. Therefore,

$$x \wedge y \equiv y \wedge y,$$

since logical equivalence is a congruence relation. But $y \wedge y$ is logically equivalent to y by the idempotence laws for conjunction (T20 and T21). Thus,

$$x \wedge y \equiv y.$$

In other words, if both x and y are theorems, then so is $x \wedge y$. Since

$$x \wedge y \leq x \quad \text{and} \quad x \wedge y \leq y,$$

the converse is also true: if $x \wedge y$ is a theorem, then so are both x and y. From these observations it follows, finally, that

$$x \equiv y \qquad \text{is equivalent to} \qquad \vdash x \Leftrightarrow y$$

(because $x \equiv y$ means, by definition, that both $x \Rightarrow y$ and $y \Rightarrow x$ are theorems). Observe that the preceding displayed line has three mutually confusable symbols, each of which could mean logical equivalence to somebody:

$$\equiv,$$

"is equivalent to",

and

$$\Leftrightarrow .$$

The first of these (despite that its official name is logical equivalence) refers to a binary relation in a certain algebraic structure; the second expresses a mathematico-logical fact; and the third is a binary operation in an algebraic structure.

18. Algebraic identities

The last set of theorems to be considered here will lead to familiar looking identities, such as the commutative, associative, and distributive laws. In terms of logical equivalence, the commutative laws for disjunction and conjunction (T3 and T17) can be formulated as follows:

$$x \vee y \equiv y \vee x,$$

$$x \wedge y \equiv y \wedge x.$$

To obtain the associative for disjunction, begin with

T22 $$\vdash x \vee (y \vee z) \Rightarrow y \vee (x \vee z).$$

For the proof, observe first that

$$\vdash z \Rightarrow x \vee z$$

(by T9), and hence that

$$\vdash y \vee z \Rightarrow y \vee (x \vee z)$$

and

$$\vdash x \vee (y \vee z) \Rightarrow x \vee (y \vee (x \vee z))$$

(by T4 and modus ponens). It follows from the commutative law for disjunction that

$$\text{(i)} \qquad \vdash x \vee (y \vee z) \Rightarrow (y \vee (x \vee z)) \vee x.$$

Next:

$$\vdash x \vee z \Rightarrow y \vee (x \vee z)$$

(by T9), whence

$$\vdash x \Rightarrow y \vee (x \vee z)$$

(by T2 and the transitivity of implication). The preceding theorem, T4, and modus ponens yield

$$\vdash (y \vee (x \vee z)) \vee x \Rightarrow (y \vee (x \vee z)) \vee (y \vee (x \vee z)),$$

and therefore (by the idempotence of $\vee$)

$$\text{(ii)} \qquad \vdash (y \vee (x \vee z)) \vee x \Rightarrow y \vee (x \vee z).$$

From (i) and (ii) the transitivity of implication yields the desired result.

The associative law for disjunction,

$$x \vee (y \vee z) \equiv (x \vee y) \vee z,$$

is now easy:

$$\begin{aligned} x \vee (y \vee z) &\equiv x \vee (z \vee y) && \text{(commutative law),} \\ &\leq z \vee (x \vee y) && \text{(T22),} \\ &\equiv (x \vee y) \vee z && \text{(commutative law),} \end{aligned}$$

and

$$(x \vee y) \vee z \equiv z \vee (y \vee x) \quad \text{(commutative law, twice),}$$

$$\leq (z \vee y) \vee x \quad \text{(by the entailment just proved)},$$

$$\equiv x \vee (y \vee z) \quad \text{(commutative law, twice)}.$$

The dual associative law for conjunction,

$$x \wedge (y \wedge z) \equiv (x \wedge y) \wedge z,$$

follows at once by the duality principle. Among many other consequences of the associative law, the following ones deserve mention:

$$x \Rightarrow (y \Rightarrow z) \equiv y \Rightarrow (x \Rightarrow z),$$

$$y \Rightarrow (x \Rightarrow z) \equiv (x \wedge y) \Rightarrow z,$$

and

$$x \Rightarrow y \equiv x \Rightarrow (x \Rightarrow y)$$

for all sentences x, y, and z. The first two of these imply that

(iii) if $x \leq (y \Rightarrow z)$, then $y \leq (x \Rightarrow z)$ and $x \wedge y \leq z$;

the last one implies that

(iv) if $x \leq (x \Rightarrow y)$, then $x \leq y$.

A related technical result (to be used once below) is that

(v) $x \leq (y \Rightarrow x \wedge y)$.

For the proof of this one, note that

$$\vdash (\neg x \vee \neg y) \vee \neg(\neg x \vee \neg y),$$

(by T7), and hence, by the associative law and the definition of conjunction,

$$\vdash \neg x \vee (\neg y \vee (x \wedge y)).$$

Now, finally, the (left-hand) distributive law for disjunction over conjunction,

$$x \vee (y \wedge z) \equiv (x \vee y) \wedge (x \vee z),$$

is within reach. Begin with the observation that

(vi) $x \vee (y \wedge z) \leq x \vee y$ and $x \vee (y \wedge z) \leq x \vee z$

(because $(y \wedge z) \leq y$ and $(y \wedge z) \leq z$—use T4). By (v) above

$$x \vee y \leq (x \vee z) \Rightarrow ((x \vee y) \wedge (x \vee z)),$$

and therefore, by (vi),

$$x \vee (y \wedge z) \leq (x \vee z) \Rightarrow ((x \vee y) \wedge (x \vee z)).$$

It follows from (iii) that

$$x \vee z \leq x \vee (y \wedge z) \Rightarrow ((x \vee y) \wedge (x \vee z)),$$

and hence, by (vi), that

$$x \vee (y \wedge z) \leq x \vee (y \wedge z) \Rightarrow ((x \vee y) \wedge (x \vee z)).$$

This yields, by (iv), that

$$x \vee (y \wedge z) \leq (x \vee y) \wedge (x \vee z),$$

which is half the distributive law.

Since

$$\begin{aligned} y &\leq (z \Rightarrow (y \wedge z)) && \text{(by (v)),} \\ &\leq (x \vee z) \Rightarrow (x \vee (y \wedge z)) && \text{(by T4),} \end{aligned}$$

it follows that

$$\begin{aligned} x \vee z &\leq (y \Rightarrow x \vee (y \wedge z)) && \text{(by (iii)),} \\ &\leq (x \vee y) \Rightarrow (x \vee (x \vee (y \wedge z))) && \text{(by T4),} \\ &\equiv (x \vee y) \Rightarrow (x \vee (y \wedge z)) \end{aligned}$$

(by the associative and idempotence laws for disjunction) and therefore, by (iii) and the commutativity of conjunction,

$$(x \vee y) \wedge (x \vee z) \leq x \vee (y \wedge z).$$

This is the second half of the distributive law.

The right-hand distributive law for disjunction over conjunction,

$$(x \wedge y) \vee z \equiv (x \vee z) \wedge (y \vee z),$$

is an immediate consequence of the left-hand law (use the commutative law for disjunction). The dual distributive laws for conjunction over disjunction,

$$x \wedge (y \vee z) \equiv (x \wedge y) \vee (x \wedge z)$$

and

$$(x \vee y) \wedge z \equiv (x \wedge z) \vee (y \wedge z),$$

now follow by the duality principle.

The absorption laws for disjunction and conjunction (T2, T9, T18, and T19) can be reformulated as algebraic identities:

$$x \wedge (x \vee y) \equiv x,$$

$$x \vee (x \wedge y) \equiv x.$$

Indeed,

$$\begin{aligned} x &\leq x \wedge x && \text{(by idempotence)},\\ &\leq x \wedge (x \vee y) && \text{(by T2)},\\ &\leq x && \text{(by T18)}, \end{aligned}$$

which yields the absorption law for disjunction. Its dual follows of course by duality.

Since $x \vee \neg x$ is always a theorem (by T7), and since everything entails a theorem, it follows that

$$y \leq x \vee \neg x,$$

and hence that

$$\neg (x \vee \neg x) \leq \neg y,$$

by the law of contraposition. Therefore

$$\neg y \vee \neg (x \vee \neg x) \leq \neg y \vee \neg y$$

(by T4), and hence

$$y \wedge y \leq y \wedge (x \vee \neg x)$$

by the law of contraposition and the definition of conjunction. Combine this with the idempotence and absorption laws for conjunction (T20 and T18) to obtain

$$y \leq y \wedge y \leq y \wedge (x \vee \neg x) \leq y.$$

This proves the identity

$$y \wedge (x \vee \neg x) \equiv y,$$

and therefore also the dual identity

$$y \vee (x \wedge \neg x) \equiv y.$$

The kind of axiom splitting that got us to this point is the least attractive and the least rewarding part of any subject. We presented it here mainly because, in this subject of all subjects, its omission would have given a distorted picture of what much of the subject is like. The result of all the manipulations is that something we were driving at all along (to be discussed soon) follows from something else. This is not at all surprising — that, after all, is the only reason why this particular something else is considered. If the initially desired result had not followed, we would happily have changed the something else. In other words: we knew the answer all along, and all that has been happening is a demonstration that the rigging of the question to produce the answer was done without a blunder. In what follows we shall usually try to avoid the axiom splitting and try to drive straight to the goal. The order in which we did things so far has been that in Hilbert & Ackermann (*Principles of Mathematical Logic*, Chelsea Publishing Company, 1950), influenced by Rosenbloom (*The Elements of Mathematical Logic*, Dover Publications, 1950).

Boolean algebra

19. Equivalence classes

There are a few more things to be said about the propositional calculus, but they are better and more clearly said in algebraic language. We proceed to develop the appropriate algebraic language, namely the language of Boolean algebra. Whenever **S** is a set and **E** is an equivalence relation between elements of **S**, it is natural to consider the $\mathbf{S}/\mathbf{E}$ (**S** modulo **E**) of equivalence classes. An *equivalence class* of **E** in **S** is a subset X of **S** such that (i) every two elements in X are equivalent (in the sense of **E**), and (ii) each element of **S** that is equivalent to some element of X belongs to X. The equivalence class *of an element* of **S** is the equivalence class to which that element belongs. If, moreover, the set **S** has an algebraic structure compatible with **E**, that is, if there are defined in **S** certain algebraic operations with the property that performing them on equivalent elements yields equivalent elements (in still other words, if **E** is a congruence relation with respect to the algebraic structure of **S**), then the quotient set $\mathbf{S}/\mathbf{E}$ inherits the algebraic structure of **S**. The purpose of what follows is to apply these remarks to the particular case where **S** is the set of sentences of the propositional calculus and **E** is the relation $\equiv$ of logical equivalence.

Let **A** be the set of all equivalence classes of sentences of the propositional calculus with respect to $\equiv$. When a generic name is needed for the elements of **A**, we shall call them *propositions*; the use of the word in this sense is not standard, but neither is it likely to outrage many scholars. Since the set of all sentences is endowed with an algebraic structure (via the operations of disjunction, conjunction, and negation), and since we have seen that logical equivalence is compatible with these operations, it makes unambiguous sense to speak of disjunction, conjunction, and negation for propositions. We shall expand on these remarks in a moment.

It is convenient at this point to change some of the (possibly tacit) alphabetic conventions. The letters p, q, and r, frozen so far as propositional variables, are hereby freed for other uses. The concept of propositional variables (in the precise sense discussed before, as certain specific "monomial" sentences of the propositional calculus) will still be needed below (once or twice), and when it is needed a temporary ad hoc notation for particular propositional variables will be established. The freed letters will be the main ones used in the discussion of **A** — elements of **A** will be denoted typically by p, q, and r. The symbols $\neg$, $\vee$, and $\wedge$, for negation, disjunction, and conjunction, will be retained. (We are using the symbol $\neg$ because it recalls the subtraction symbol that was used historically for the Boolean operation corresponding to negation. Another symbol that is sometimes used for this purpose is $'$.)

Suppose now that p is an element of **A** and that $x \in p$, and write $\neg p$ for the equivalence class of $\neg x$. The element $\neg p$ of **A** is unambiguously determined by p and is independent of the choice of x: if both x_1 and x_2 are in the equivalence class of p, so that $x_1 \equiv x_2$, then $\neg x_1 \equiv \neg x_2$, so that $\neg x_1$ and $\neg x_2$ belong to the same equivalence class. Similarly, if p and q are elements of **A**, and if $x \in p$ and $y \in q$, write $p \vee q$ for the equivalence class of $x \vee y$, and $p \wedge q$ for the equivalence class of $x \wedge y$. The compatibility of logical equivalence and the propositional connectives (discussed earlier) implies that the elements $p \vee q$ and $p \wedge q$ are unambiguously determined by p and q (and are independent of the choices of x and y). The most important fact about **A** and the operations $\neg$, $\vee$, and $\wedge$ is that whenever p, q, and r are in **A**, then

B1 $p \vee q = q \vee p$ $\qquad$ $p \wedge q = q \wedge p$

B2 $p \vee (q \vee r) = (p \vee q) \vee r$ $\qquad$ $p \wedge (q \wedge r) = (p \wedge q) \wedge r$

B3 $p \vee (p \wedge q) = p$ $\qquad$ $p \wedge (p \vee q) = p$

B4 $p \vee (q \wedge r) = (p \vee q) \wedge (p \vee r)$ $\qquad$ $p \wedge (q \vee r) = (p \wedge q) \vee (p \wedge r)$

B5 $p \vee (q \wedge \neg q) = p$ $\qquad$ $p \wedge (q \vee \neg q) = p.$

All of these identities were proved in section 18. Thus, B1 and B2 are the commutative and associative laws for disjunction and conjunction, B3 the absorption laws for conjunction and disjunction, and B4 the distributive laws for disjunction over conjunction and for conjunction over disjunction.

The importance of these identities is that they serve to characterize (and hence can be used to define) a fundamental algebraic concept. A *Boolean algebra* is a non-empty set, such as **A**, with a distinguished unary operation

$\neg$, and two distinguished binary operations $\vee$ and $\wedge$, satisfying the identities B1–B5. In the context of Boolean algebra, the operations $\vee$ and $\wedge$ are sometimes referred to as *join* and *meet* respectively, while $\neg$ is called *complementation*. If a Boolean algebra has just one element, then that algebra is called *trivial*, and *non-trivial* otherwise. The propositions (not sentences) of the propositional calculus form a Boolean algebra $\mathbf{A}$ with respect to disjunction, conjunction, and negation. If we knew that $\mathbf{A}$ contained at least two distinct elements, then we could conclude that it is non-trivial. This desideratum turns out to be somewhat more complicated to treat than might appear from a casual glance.

20. Interpretations

The problem has to do with the concept of interpretation. An *interpretation* of the propositional calculus is a mapping φ that associates with each sentence one of the two numbers 0 and 1 in such a way that

$$\varphi(x \vee y) = \max(\varphi(x), \varphi(y))$$

and

$$\varphi(\neg x) = 1 - \varphi(x).$$

(This definition of interpretation can, should, and will undergo a sweeping generalization later.) The motivation has to do with truth and falsity. Think of 1 as truth and 0 as falsity. With this interpretation of the symbols, an interpretation becomes an assignment of truth values to each sentence in an algebraically (logically) coherent manner. The latter qualification means that the assignment of truth values should be such that the disjunction of two sentences is false if and only if both are, and the negation of a sentence is false if and only if the sentence itself is true. This qualification is exactly what the algebraically expressed conditions on φ say.

Are there any interpretations? For our present purposes a simple affirmative answer would be useful, but it turns out that it is no more difficult to prove that there are many interpretations than it is to prove that there are any. Recall, therefore, that among the sentences of the propositional calculus there was a distinguished (countably infinite) set $\mathbf{V}$ of sentences called propositional variables (and then denoted by p, p^{+}, p^{++}, p^{+++}, etc.). Suppose now that φ_0 is an arbitrary mapping of $\mathbf{V}$ into the two-element set $\{0, 1\}$; we assert that φ_0 can be extended in one and only one way so as to become an interpretation φ of the propositional calculus.

The details of the extension are a bit of mechanical fuss with symbolism. The idea is easy enough. Every sentence of the propositional calculus is made up of elements of **V** and the result of operating on them by the propositional connectives $\neg$ and $\vee$. It follows that every sentence has either exactly one "immediate ancestor" (in case it has the form $\neg x$, where x is some shorter sentence), or exactly two immediate ancestors (in case it has the form $x \vee y$, where x and y are shorter sentences). The process of defining φ is inductive. Once $\varphi(x)$ is known, $\varphi(\neg x)$ is *defined* as $1 - \varphi(x)$, and once $\varphi(x)$ and $\varphi(y)$ are known, $\varphi(x \vee y)$ is defined as $\max(\varphi(x), \varphi(y))$. As an example, consider the sentence

$$((p \vee \neg q) \vee r) \vee (\neg (p \vee (\neg r \vee q))).$$

(Here we take p, q, and r to be elements of **V**.) If

$$\varphi_0(p) = 1, \quad \varphi_0(q) = 0, \quad \text{and} \quad \varphi_0(r) = 1,$$

then $\varphi(\neg q) = 1$, so

$$\varphi(p \vee \neg q) = \max(\varphi(p), \varphi(\neg q)) = \max(1, 1) = 1,$$

and therefore

$$\varphi((p \vee \neg q) \vee r) = \max(\varphi(p \vee \neg q), \varphi(r)) = \max(1, 1) = 1.$$

Also, $\varphi(\neg r) = 0$, so $\varphi(\neg r \vee q) = 0$, and therefore $\varphi(p \vee (\neg r \vee q))$ must be 1. It follows that $\varphi(\neg (p \vee (\neg r \vee q)))$ must be 0, and hence that the value of φ at the displayed sentence must be 1. Note that once $\varphi((p \vee \neg q) \vee r)$ is known to be 1, further computations are unnecessary; the value of φ at the entire displayed sentence must be 1, regardless of what the other constituent contributes. This sort of procedure of deriving the value of φ from prescribed values of φ_0 is usually done in terms of "truth tables".

It is an immediate consequence of the extension statement whose proof was just indicated that not only do interpretations exist, but, in fact, they exist in large numbers — there are infinitely many (and in fact continuum many) of them. Once the existence of interpretations (and in fact of interpretations that extend arbitrary truth-value assignments to the propositional variables) is granted, it follows easily that not every sentence is a theorem (and hence that there are at least two distinct logical equivalence classes of sentences). The reason is that every interpretation assigns the value 1 (truth) to every theorem. A sentence to which every interpretation assigns the value 1 is said to be *valid*.

Valid sentences are called *tautologies*. Thus, every theorem is a tautology. (In different words, the propositional calculus is *semantically sound*.)

The proof of this for the axioms is a straightforward check, which it is sufficient to illustrate with axiom schemata T1 and T2. Thus: if φ is an interpretation and if x is a sentence, then

$$\varphi(x \vee x) = \varphi(x),$$

hence

$$\varphi(\neg(x \vee x)) = 1 - \varphi(x),$$

and therefore

$$\begin{aligned}\varphi((x \vee x) \Rightarrow x) &= \varphi(\neg(x \vee x) \vee x)\\ &= \max(1 - \varphi(x), \varphi(x)) = 1.\end{aligned}$$

Also: if φ is an interpretation and x and y are sentences, then

$$\begin{aligned}\varphi(x \Rightarrow (x \vee y)) &= \varphi(\neg x \vee (x \vee y))\\ &= \max(1 - \varphi(x), \max(\varphi(x), \varphi(y)))\\ &\geq \max(1 - \varphi(x), \varphi(x)) = 1.\end{aligned}$$

If it is known that $\varphi(x) = 1$ whenever x is an axiom, it remains only to prove that modus ponens leads from "true" sentences to "true" ones only. That is, it is to be proved that if $\varphi(x) = 1$ and $\varphi(x \Rightarrow y) = 1$, then $\varphi(y) = 1$. Suppose $\varphi(x) = 1$ and $\varphi(x \Rightarrow y) = 1$. Then

$$\begin{aligned}1 = \varphi(x \Rightarrow y) &= \varphi(\neg x \vee y)\\ &= \max(1 - \varphi(x), \varphi(y))\\ &= \max(0, \varphi(y)) = \varphi(y).\end{aligned}$$

Consequence: not every sentence is provable (because if x is a theorem, then $\varphi(\neg x) = 0$).

21. Consistency and Boolean algebra

In algebraic terms the conclusion we have reached is, as advertised, that the set of all propositions of the propositional calculus is a non-trivial Boolean algebra. In logical terms, the conclusion is that the propositional calculus is *consistent*. An intuitive justification for the use of the word goes as follows. If there were only one equivalence class, then any two sentences would be logically equivalent to each other and, in particular, to all theorems. Because a sentence that is logically equivalent to a theorem must itself be a theorem (by modus ponens), every sentence would be provable. In particular, for each sentence x, both x and $\neg x$ would be provable. This corresponds to the intuitive concept of a contradiction — inconsistency.

If, on the other hand, there is more than one equivalence class, then it can never happen that both x and $\neg x$ are provable. Reason: if x and $\neg x$ are provable, then so is $x \wedge \neg x$. Furthermore, $x \vee \neg x$ is always provable, by T7, and therefore so is $(x \vee \neg x) \vee y$ (for every sentence y), by T2 and modus ponens. Since

$$\begin{aligned}
(x \vee \neg x) \vee y &\equiv \neg\neg(x \vee \neg x) \vee y && \text{(double negation)},\\
&\equiv \neg(x \vee \neg x) \Rightarrow y && \text{(definition of } \Rightarrow),\\
&\equiv \neg(\neg x \vee x) \Rightarrow y && \text{(commutativity)},\\
&\equiv \neg(\neg x \vee \neg\neg x) \Rightarrow y && \text{(double negation)},\\
&\equiv (x \wedge \neg x) \Rightarrow y && \text{(definition of } \wedge),
\end{aligned}$$

it follows that $(x \wedge \neg x) \Rightarrow y$ is provable, and therefore (by modus ponens) that every y is provable. Since any two theorems are equivalent, there can only be one equivalence class. Conclusion: the intuitive idea of inconsistency coincides with the assertion that any two sentences are provably equivalent to one another, or what amounts to the same thing, that every sentence is provable.

22. Duality and commutativity

Let us now make explicit and official something that was discussed in the context of the propositional calculus, namely that $\vee$ and $\wedge$ play perfectly symmetric roles in the axioms Bl–B5. If they are interchanged throughout, the axioms (as a whole) remain unchanged. From this it follows that if, in any statement about the elements of a Boolean algebra, $\vee$ and $\wedge$ are interchanged, the resulting statement and the original one are true or false at the same time. This fact is usually referred to as the *principle of duality*. The general theorems about Boolean algebras, and, for that matter, their proofs also, come in dual pairs. A practical consequence of this principle, often exploited in the sequel, is that in the theory of Boolean algebras it is sufficient to state and to prove only half the theorems; the other half come gratis from the principle of duality.

A slight misunderstanding can arise about the meaning of duality, and sometimes does. It is well worth while to clear it up once and for all, especially since the clarification is quite amusing in its own right. If an experienced Boolean algebraist is asked for the dual of a Boolean polynomial, such as say $p \vee q$, the answer might be $p \wedge q$ one day and $\neg p \wedge \neg q$ another day; the answer $\neg p \vee \neg q$ is less likely but not impossible. (The definition of Boolean polynomials is the same as that of ordinary polynomials, except the admissible operations are not addition and multiplication but join (that is $\vee$), meet (that is $\wedge$), and complementation (that is $\neg$). What is needed here to avoid confusion is a little careful terminological distinction. Let us restrict attention to the completely typical case of a polynomial $f(p,q)$ in two variables. The *complement* of $f(p,q)$ is by definition $\neg f(p,q)$; the *dual* of $f(p,q)$ is defined to be

$$\neg f(\neg p, \neg q);$$

the *contradual* of $f(p,q)$ is defined to be

$$f(\neg p, \neg q).$$

In the case of the polynomial $p \vee q$, the complement $\neg(p \vee q)$ is equal to $\neg p \wedge \neg q$, and the dual $\neg(\neg p \vee \neg q)$ is equal to $p \wedge q$; the contradual is $\neg p \vee \neg q$.

What goes on here is that there is a group acting on the set of propositions, and this group is of order four, not two. Specifically, consider the four mappings that take any given proposition to (a) itself, (b) its complement, (c) its dual, and (d) its contradual. Each of these mappings is a permutation of the set of propositions, and the set of these mappings forms a group under the operation of functional composition. The group is the Klein four-group. This comment was explicitly made in print by Walter Gottschalk; he describes the situation by speaking of the principle of quaternality.

Once a Boolean identity has been proved, it follows immediately from B1 that "commuted" versions of the identity also hold. For example, the right-hand distributive laws,

$$(p \wedge q) \vee r = (p \vee r) \wedge (q \vee r),$$

$$(p \vee q) \wedge r = (p \wedge r) \vee (q \wedge r),$$

are an immediate consequence of the left-hand distributive laws, B4 and B1. Once a Boolean law has been established, commuted versions can and will be used freely, without formulating them explicitly.

23. Properties of Boolean algebras

Let us now return to the axioms of Boolean algebras and derive some of their consequences. (Warning: in this enterprise we shall operate with B1–B5 only, and not with the properties of the propositional calculus that gave rise to B1–B5. In other words, we are about to study the general concept of a Boolean algebra, not the special concept of the Boolean algebra induced by the propositional calculus. Somewhat later we shall examine the relation between the special concept and the generalization, and the extent to which the latter generalizes the former.)

An easy consequence of the axioms is that both $\wedge$ and $\vee$ are idempotent:

$$p \vee p = p \qquad \text{and} \qquad p \wedge p = p.$$

Proof.

$$\begin{aligned} p &= p \vee (p \wedge q) && \text{(by B3)},\\ &= (p \vee p) \wedge (p \vee q) && \text{(by B4)},\\ &= (p \wedge (p \vee q)) \vee (p \wedge (p \vee q)) && \text{(by B4)},\\ &= p \vee p && \text{(by B3)}. \end{aligned}$$

The proof of the dual law $p \wedge p = p$ is the dual of the above proof.

There is a natural (partial) order relation associated with every Boolean algebra; the relation

$$p \leq q$$

is defined to mean

$$p \wedge q = p,$$

or, equivalently,

$$p \vee q = q.$$

(The equivalence of these two equations follows from the absorption laws, B3.)

Let us proceed to verify that $\leq$ is indeed a partial order. The reflexivity of order ($p \leq p$) follows from the idempotence of $\wedge$ (or $\vee$). If $p \leq q$ and $q \leq p$, so that

$$p \wedge q = p \quad \text{and} \quad q \wedge p = q,$$

then the commutative law implies that $p = q$, so that $\leq$ is antisymmetric. If, finally,

$$p \leq q \quad \text{and} \quad q \leq r,$$

then

$$p \wedge q = p \quad \text{and} \quad q \wedge r = q,$$

so that

$$p = p \wedge q = p \wedge (q \wedge r) = (p \wedge q) \wedge r = p \wedge r,$$

and therefore $p \leq r$.

In the language of order, B5 (with p and q interchanged) says that

$$p \wedge \neg p \leq q \quad \text{and} \quad q \leq p \vee \neg p$$

for all p and q. Replace q by $q \wedge \neg q$ in the first of these inequalities and by $q \vee \neg q$ in the second to get

$$p \wedge \neg p \leq q \wedge \neg q \quad \text{and} \quad q \vee \neg q \leq p \vee \neg p$$

for all p and q. Then interchange p and q and use the antisymmetry of order to infer that

$$p \wedge \neg p = q \wedge \neg q \quad \text{and} \quad q \vee \neg q = p \vee \neg p$$

for all p and q. It is customary to write

$$0$$

for the constant value of the Boolean "polynomial" $p \wedge \neg p$ and

$$1$$

for that of

$$p \vee \neg p.$$

In this notation B5 can be written as

$$p \vee 0 = p \qquad p \wedge 1 = p.$$

In words, 0 and 1 are the identity elements for the operations $\vee$ and $\wedge$ respectively. As is clear from their definitions, the elements 0 and 1 are the duals of one another.

Two related laws are

$$p \wedge 0 = 0 \quad \text{and} \quad p \vee 1 = 1.$$

Proof.

$$\begin{aligned} p \wedge 0 &= p \wedge (p \wedge \neg p) && \text{(by definition of 0)}, \\ &= (p \wedge p) \wedge \neg p && \text{(by B2)}, \\ &= p \wedge \neg p && \text{(by idempotence)}, \\ &= 0 && \text{(by definition of 0)}. \end{aligned}$$

The second law follows by duality. Note that B5 can be expressed in the form

$$0 \leq p \leq 1$$

for all p.

A moment's time out to make contact with the propositional calculus. Note that if p and q are the equivalence classes of the sentences x and y, then $p \leq q$ happens if and only if $x \leq y$. The equivalence class 1 is the set of theorems; the equivalence class 0 is the set of all those sentences whose negations are theorems.

There is still more routine work to be done. Here is a small theorem, the *monotony law* for $\vee$:

$$\text{if} \quad p \leq q \quad \text{and} \quad r \leq s, \quad \text{then} \quad p \vee r \leq q \vee s.$$

Indeed: if $p \vee q = q$ and $r \vee s = s$, then

$$(p \vee r) \vee (q \vee s) = q \vee s.$$

Duality implies the monotony law for $\wedge$:

$$\text{if} \quad p \leq q \quad \text{and} \quad r \leq s, \quad \text{then} \quad p \wedge r \leq q \wedge s.$$

It follows that

$$\text{if} \quad p \leq q \quad \text{and} \quad r \leq q, \quad \text{then} \quad p \vee r \leq q,$$

and

$$\text{if} \quad p \leq q \quad \text{and} \quad p \leq s, \quad \text{then} \quad p \leq q \wedge s.$$

Also:

$$p \leq p \vee q \quad \text{and} \quad q \leq p \vee q,$$

since

$$p \vee (p \vee q) = (p \vee p) \vee q = p \vee q;$$

by duality,

$$p \wedge q \leq p \quad \text{and} \quad p \wedge q \leq q.$$

A useful rephrasing of all these facts is that $p \vee q$ and $p \wedge q$ can be defined in terms of $\leq$; indeed, $p \vee q$ is the least element of $\mathbf{A}$ that dominates both p and q (and $p \wedge q$ is the greatest element of $\mathbf{A}$ that is dominated by both p and q). For this reason $p \vee q$ is often called the *supremum* (sup) or least upper bound (l.u.b.) of p and q, and, similarly, $p \wedge q$ is called their *infimum* (inf) or greatest lower bound (g. l. b).

Complementation is completely characterized by the definitions of 0 and 1:

$$\text{if} \quad p \wedge q = 0 \quad \text{and} \quad p \vee q = 1, \quad \text{then} \quad q = \neg p.$$

Indeed, if $p \vee q = 1$, then

$$\begin{aligned} q &= 0 \vee q \\ &= (p \wedge \neg p) \vee q \\ &= (p \vee q) \wedge (\neg p \vee q) \\ &= 1 \wedge (\neg p \vee q) \\ &= \neg p \vee q, \end{aligned}$$

so that

$$\neg p \leq q.$$

On the other hand, if $p \wedge q = 0$, then

$$\begin{aligned} q &= 1 \wedge q \\ &= (p \vee \neg p) \wedge q \\ &= (p \wedge q) \vee (\neg p \wedge q) \end{aligned}$$

$$= 0 \vee (\neg p \wedge q)$$

$$= \neg p \wedge q,$$

so that $q \leq \neg p$.

Since $\neg p \wedge p = 0$ and $\neg p \vee p = 1$ for all p, it follows from the characterization of complements that

$$\neg \neg p = p$$

(recall the law of double negation). Consequence:

$$\text{if } \neg p = \neg q, \quad \text{then} \quad p = \neg \neg p = \neg \neg q = q.$$

In other words, the function that takes each element to its complement is a bijection of the elements of the Boolean algebra.

The De Morgan laws,

$$\neg (p \vee q) = \neg p \wedge \neg q,$$

$$\neg (p \wedge q) = \neg p \vee \neg q,$$

also follow easily now: if $r = \neg p \wedge \neg q$, then

$$\begin{aligned} (p \vee q) \wedge r &= (p \wedge (\neg p \wedge \neg q)) \vee (q \wedge (\neg p \wedge \neg q)) \\ &= ((p \wedge \neg p) \wedge \neg q) \vee ((q \wedge \neg q) \wedge \neg p) \\ &= 0 \vee 0 = 0 \end{aligned}$$

and

$$\begin{aligned} (p \vee q) \vee r &= ((p \vee q) \vee \neg p) \wedge ((p \vee q) \vee \neg q) \\ &= ((p \vee \neg p) \vee q) \wedge (p \vee (q \vee \neg q)) \\ &= 1 \wedge 1 = 1, \end{aligned}$$

whence $r = \neg (p \vee q)$. The dual law follows from duality.

The observations of the last two paragraphs can be summarized algebraically as follows. Let **B** be the Boolean algebra with operations $\vee$, $\wedge$, and $\neg$, and **B*** the algebra obtained from **B** by interchanging $\vee$ and $\wedge$. (Thus, $\wedge$ becomes the join, and $\vee$ the meet, of **B***.) Then the function that takes p to $\neg p$ for every p in **B** is an isomorphism between the two algebras.

It follows that $p \leq q$ if and only if $\neg q \leq \neg p$ (since $p \vee q = q$ if and only if $\neg p \wedge \neg q = \neg q$). It follows also that

$$p \vee q = \neg(\neg p \wedge \neg q)$$

and

$$p \wedge q = \neg(\neg p \vee \neg q),$$

so that either one of meet and join can be defined in terms of the other (and complementation). Put $q = \neg p$ in these relations to get $1 = \neg 0$ and $\neg 1 = 0$.

24. Subtraction

We conclude this study of the elements of a Boolean algebra by introducing some more operations and another relation pertinent to Boolean algebras. The first operation is called *subtraction*.

Define $p - q$ to be $p \wedge \neg q$, and note that it follows that

$$p \leq q \quad \text{if and only if} \quad p - q = 0.$$

(Argument: if $p \leq q$, then

$$p \wedge \neg q = (p \wedge q) \wedge \neg q = p \wedge (q \wedge \neg q) = p \wedge 0 = 0.$$

If $p \wedge \neg q = 0$, then

$$\begin{aligned} p = p \wedge 1 &= p \wedge (q \vee \neg q) \\ &= (p \wedge q) \vee (p \wedge \neg q) \\ &= (p \wedge q) \vee 0 = p \wedge q, \end{aligned}$$

and hence $p \leq q$.)

Since

$$\neg(p - q) = \neg(p \wedge \neg q) = \neg p \vee q,$$

it follows that $p \leq q$ if and only if $\neg p \vee q = 1$.

There are other operations of interest in Boolean algebras. For example, a new binary operation $\Rightarrow$ can be obtained by defining $p \Rightarrow q$ to be $\neg p \vee q$.

Still another: $\Leftrightarrow$ is defined by taking $p \Leftrightarrow q$ to be

$$(q \Rightarrow p) \wedge (p \Rightarrow q).$$

Note that the dual of $p - q$ is $p \vee \neg q$, which is just $q \Rightarrow p$.

The operation of *symmetric difference* (or Boolean sum) is defined by

$$p + q = (p \wedge \neg q) \vee (\neg p \wedge q).$$

For those who know what a *ring* (with unit) is, in the ordinary sense of algebra, it is easy to verify that a Boolean algebra becomes a ring with respect to this addition and the operation of multiplication defined by

$$pq = p \wedge q.$$

In this ring, the identity element for addition is 0:

$$p + 0 = (p \wedge \neg 0) \vee (\neg p \wedge 0) = (p \wedge 1) \vee 0 = p.$$

The identity element for multiplication is easily seen to be 1. Also,

$$p + p = 0,$$

that is, the *characteristic* of the ring is 2. Notice that the operation of complementation can be recovered from the ring operations:

$$\neg p = p + 1.$$

Since join is definable in terms of complementation and multiplication, the original Boolean operations can be recovered from the ring operations.

The dual of the Boolean sum $p + q$ is, by definition, the polynomial

$$(p \vee \neg q) \wedge (\neg p \vee q) = (q \Rightarrow p) \wedge (p \Rightarrow q) = (p \Leftrightarrow q).$$

Another well-known Boolean operation that deserves mention is the *Sheffer stroke*; it is defined by

$$p \mid q = \neg p \wedge \neg q = \neg (p \vee q).$$

In logical contexts this operation is sometimes known as *binary rejection* (neither p nor q). In analogy with the use of the words "and" and "or" in logic, it is sometimes referred to as "nor". The chief theoretical application of the Sheffer stroke is the remark that a single operation, namely the stroke, is enough to define Boolean algebras. To prove this, it is sufficient to show that complementation, meet, and join can be expressed in terms of the stroke, and, indeed,

$$\neg p = p \mid p,$$

$$p \wedge q = (p \mid p) \mid (q \mid q),$$

and

$$p \vee q = (p \mid q) \mid (p \mid q).$$

We see thus that the process of defining many useful operations and relations in a Boolean algebra is to some extent reversible. This fact is responsible for the plethora of different axiomatic approaches to the subject. A Boolean algebra can be defined in terms of its partial order, or in terms of complements and meets, or in terms of complements and joins, or in ring-theoretic terms (addition and multiplication), and so on and so forth *ad* almost *infinitum*.

Define two elements p and q to be *disjoint* if $p \wedge q = 0$. Thus p and $q - p$ are always disjoint, because

$$p \wedge (q \wedge \neg p) = q \wedge 0 = 0.$$

Since

$$\begin{aligned} p \vee (q \wedge \neg p) &= (p \vee q) \wedge (p \vee \neg p) \\ &= (p \vee q) \wedge 1 \\ &= p \vee q, \end{aligned}$$

it follows that

$$p \vee (q - p) = p \vee q.$$

In sum, the elements p and $q - p$ form a *partition* of $p \vee q$.

25. Examples of Boolean algebras

The easiest example of a non-trivial Boolean algebra is not the one that was derived from the propositional calculus, but the two-element set $\{0, 1\}$, with $\neg p, p \vee q$, and $p \wedge q$ defined (in terms of familiar arithmetic operations) to be $1 - p, \max(p, q)$, and $p \cdot q$ respectively. We shall follow the common practice of denoting this algebra by **2**.

Another easy class of examples is obtained this way: let X be an arbitrary non-empty set, and let $\mathbf{A}$ be the collection of all subsets of X. For subsets P and Q of X (the use of capital letters is traditional in such contexts), if $\neg P$, $P \vee Q$, and $P \wedge Q$ are defined to be the (set-theoretic) complement of P (in X), the union of P and Q, and the intersection of P and Q, the set $\mathbf{A}$ becomes a Boolean algebra. (Note that the example of the algebra $\mathbf{2}$ is a special case: take X to be a set with just one element, which shall remain nameless. The resulting algebra of sets is essentially the same as $\mathbf{2}$: X plays the role of 1, and the empty set — usually denoted by $\varnothing$ — plays the role of 0.)

For another example, take the set X of all positive integers, and define a collection $\mathbf{A}$ of subsets of X as follows: a set P belongs to $\mathbf{A}$ if and only if P is finite or co-finite, that is, either P or the complement of P (in X) is finite. With the same operations as before (complementation, union, intersection), the set $\mathbf{A}$ is a Boolean algebra. This example breeds many others. If the set of positive integers is replaced by any other set X, another example results. If the set X is finite, the example is not new (in that case every subset is finite, so we just end up with the collection of all subsets of X). If the set X is countably infinite, the example differs from the one associated with the positive integers in notation only. If the cardinality (that is, the "size") of X is allowed to vary over uncountable cardinals, new examples result.

Another direction of generalization is to replace the cardinal inequality implied by the use of the word "finite" (that is, the inequality "card $P < \aleph_0$") by others. Thus, for any set X and for any infinite cardinal $\aleph$, let $\mathbf{A}$ be the collection of all those subsets P of X for which either card $P < \aleph$ or card $\neg P < \aleph$; the result is a Boolean algebra. Example: for subsets of the set X of all real numbers, let $\mathbf{A}$ be defined as follows: a subset P of X belongs to $\mathbf{A}$ if and only if P is either countable or co-countable, that is, either P or its complement is countable.

All these examples are called *fields* of sets (not to be confused with the use of the word in another part of algebra). A field of sets is a collection of subsets of a set closed under the formation of the complement of a set and the union and intersection of two sets.

For another example (of a field of sets) let X be the set of all integers (positive, negative, or zero), and let m be an arbitrary integer. A subset P of X is called *periodic* of *period* m if it coincides with the set obtained by adding m to each of its elements. The class $\mathbf{A}$ of all periodic sets of period m is a field of subsets of X. If $m = 0$, then $\mathbf{A}$ is simply the collection of all subsets of X. If $m = 1$, then $\mathbf{A}$ consists of just the two sets $\varnothing$ and X. More generally, if m

is any positive integer, then **A** consists of all possible unions of equivalence classes of the integers modulo m. It is not difficult to check that in this case **A** is essentially the same as the field of subsets of a finite set with m elements. (In technical language, the two Boolean algebras are *isomorphic*.)

Next, let X be the set of all real numbers. A *left half-closed interval* (or, for brevity, since this is the only kind we shall consider, a half-closed interval) is a set of one of the forms $(-\infty, b)$, or $[a, b)$, or $[a, +\infty)$, that is, the set of all those elements x in X for which $x < b$, or $a \leq x < b$, or $a \leq x$, where, of course, a and b themselves are real numbers and $a < b$. The class **A** of all finite unions of half-closed intervals is a field of subsets of X. A useful variant of this example uses the closed unit interval $[0, 1]$ in the role of X. In that case it is convenient to stretch the terminology so as to include the closed intervals $[a, 1]$ and the degenerate interval $[1]$ among the half-closed intervals.

Valuable examples of fields of sets can be defined in the plane, as follows. Call a subset P of the plane *vertical* if, along with each point of P, every point of the vertical line through that point also belongs to P. In other words, P is vertical if the presence of (x_0, y_0) in P implies the presence in P of (x_0, y) for every y. If **A** is any field of subsets of the plane, then the class of all vertical sets in **A** is another, and, in particular, the class of all vertical sets is a field of sets. Here are two comments that are trivial but sometimes useful: (i) the *horizontal* sets (whose definition may safely be left to the reader) constitute just as good a field as the vertical sets, and (ii) the Cartesian product of any two non-empty sets is, for these purposes, just as good as the plane.

All the examples of Boolean algebras so far have been fields of sets. Here is an example that, at first glance, looks different. Let m be an integer greater than 1, and let **A** be the set of all positive integral divisors of m. Define $\neg p$, $p \vee q$, and $p \wedge q$ by

$$\neg p = \frac{m}{p},$$

$$p \vee q = \text{l.c.m.}(p, q),$$

$$p \wedge q = \text{g.c.d.}(p, q).$$

For some integers m the result is a Boolean algebra and for others it is not. It turns out, in fact, that **A** is a Boolean algebra if and only if m is square-free (that is, m is not divisible by the square of any prime). Indeed, suppose that

$$m = m_1^{\alpha_1} \cdots m_k^{\alpha_k},$$

where the m_i's are primes and the α_i's are positive integers. Consider first the case where m is not square-free, so that, say, $\alpha_i > 1$. If $p = m_i$, then $\neg p$ is divisible by $m_i^{\alpha_i - 1}$, which, in turn, is divisible by m_i, so that we get the preposterous result $p \leq \neg p$. (Caution: this is preposterous only if p is not the zero element of the Boolean algebra. For this example the roles of the Boolean zero and unit are played by the integers 1 and m respectively.) If, on the other hand, m is square-free, then there is a natural correspondence between the subsets of $\{m_1, \ldots, m_k\}$ and the divisors of m: each subset corresponds to the product of its elements. It is easy to verify that under this correspondence the algebraic operations, as defined above, correspond to the usual set-theoretic ones. The distinction between this algebra and a certain field of sets is illusory and merely notational: the two algebras are isomorphic. We shall see a little later that this is not accidental: every Boolean algebra is isomorphic to some field of sets.

Boolean universal algebra

26. Subalgebras

The structure of Boolean algebras is in many respects typical of algebra in general. There are several universal concepts (subalgebra, homomorphism, ideal, etc.) that must be understood, and we turn now to the examination of those concepts.

A *Boolean subalgebra* of a Boolean algebra **A** is a subset **B** of **A** such that **B**, with respect to the operations of **A**, is a Boolean algebra. Since a Boolean algebra must, by definition, contain at least one element, a subalgebra can never be empty. If **B** is a subalgebra of **A** and if $p \in \mathbf{B}$, then $\neg p \in \mathbf{B}$ and therefore both 0 ($= p \wedge \neg p$) and 1 ($= p \vee \neg p$) are in **B**. These elementary observations can be used as a warning sign to avoid an obvious mistake. If **A** is the collection of all subsets of a set X, if Y is a proper subset of X, and if **B** is the collection of all subsets of Y, then **B** is *not* a subalgebra of **A**. Reason: the unit element of **A** (that is, 1) is X, which is not contained in **B**. There is another possible source of misunderstanding, but one that is less likely to lead to error. (Reason: it is not special to Boolean algebras, but has its analogue in every algebraic structure.) To be a Boolean subalgebra it is not enough to be a subset that is a Boolean algebra in its own right, however natural the Boolean operations may appear. The Boolean operations of a subalgebra must, by definition, be the restrictions of the Boolean operations of the whole algebra.

Every Boolean algebra **A** includes the subalgebra $\{0, 1\}$ of constants (which are distinct unless **A** is trivial). Every Boolean algebra **A** includes an *improper* subalgebra, namely **A** itself; all other subalgebras will be called *proper*.

If a non-empty subset **B** of a Boolean algebra **A** is closed under some Boolean operations, and if there are enough of those operations that all other Boolean operations can be defined by them, then **B** is a subalgebra of **A**.

Example: if **B** is closed under joins and complements, then **B** is a subalgebra; alternatively, if **B** is closed under the Sheffer stroke, then **B** is a subalgebra.

A moment's thought shows that the intersection of every collection of subalgebras of a Boolean algebra **A** is again a subalgebra of **A**. It follows that if **E** is an arbitrary subset of **A**, then the intersection of all those subalgebras that happen to include **E** is a subalgebra. (There is always at least one subalgebra of **A** that includes **E**, namely the improper subalgebra **A**.) That intersection, say **B**, is the smallest subalgebra of **A** that includes **E**. The algebra **B** is called the subalgebra *generated* by **E**. Thus, for example, if **E** is empty and **A** is non-trivial, then the subalgebra generated by **E** is the smallest possible subalgebra of **A**, namely **2**. A generating subset **E** of a subalgebra **B** is also known as a set of *generators* of **B**.

If the generating set **E** of **B** is not empty, then it generates **B** in a quite concrete way. To describe this generation process, let us introduce the following notation: for any subset X of **A** write

$$X^* = \{p \vee q : p, q \in X\} \cup \{\neg p : p \in X\}.$$

Put $B_0 = \mathbf{E}$; if B_n is defined, put

$$B_{n+1} = B_n^* \cup B_n.$$

Then **B** is the union of the sets B_n. Indeed, it is easy to check that this union is a subalgebra of **A** that includes **E**. Therefore, it includes **B**. On the other hand, an easy induction on n shows that B_n is included in **B** for each n, so the union of these sets is included in **B**.

27. Homomorphisms

Perhaps the most important algebraic concept is that of a homomorphism. A *Boolean homomorphism* is a mapping f from a Boolean algebra **B**, say, to a Boolean algebra **A**, such that

(i) $$f(p \wedge q) = f(p) \wedge f(q),$$

(ii) $$f(p \vee q) = f(p) \vee f(q),$$

(iii) $$f(\neg p) = \neg f(p),$$

whenever p and q are in **B**. In a somewhat loose but brief and suggestive phrase, a homomorphism is a *structure-preserving* mapping between

Boolean algebras. A convenient synonym for "homomorphism from **B** to **A**" is "**A**-valued homomorphism on **B**". Such expressions will be used most frequently in case $\mathbf{A} = \mathbf{2}$. Sometimes we shall indicate that f is a homomorphism from **B** to **A** by writing

$$\mathbf{B} \underset{f}{\rightarrow} \mathbf{A}.$$

Special kinds of Boolean homomorphisms may be described in the same words as are used elsewhere in algebra. A homomorphism may be one-to-one into (*monomorphism*: if $f(p) = f(q)$, then $p = q$); it may be onto (*epimorphism*: every element of **A** is equal to $f(p)$ for some p in **B**); it may be both one-to-one and onto (*isomorphism*); its range may be included in its domain (*endomorphism*: $\mathbf{A} \subset \mathbf{B}$); and it may be a one-to-one mapping of its domain onto itself (*automorphism*). If there exists an isomorphism from **B** onto **A**, then **A** and **B** are called *isomorphic*.

The distinguished elements 0 and 1 play a special role for homomorphisms just as they do for subalgebras. Indeed, if f is a Boolean homomorphism and if p is an element in its domain ($p = 0$ will do), then

$$f(p \wedge \neg p) = f(p) \wedge \neg f(p),$$

and therefore

(iv) $$f(0) = 0.$$

The dual argument proves the dual fact

(v) $$f(1) = 1.$$

Thus, the mapping that sends every element of one Boolean algebra onto the zero element of another is a homomorphism only if the algebra that is being mapped into is trivial. The equations (iv) and (v) imply that 0 and 1 belong to the range of every homomorphism; a glance at the equations (i)–(iii) should complete the proof that the range of every homomorphism, from **B** into **A**, say, is a Boolean subalgebra of **A**. The range of a homomorphism with domain **B** is called a *homomorphic image* of **B**.

Since every Boolean operation (for example $+$ and $\Rightarrow$) can be defined in terms of $\vee$, $\wedge$, and $\neg$ by "polynomials", it follows that a Boolean homomorphism preserves all such operations. If, that is, f is a Boolean homomorphism and p and q are elements of its domain, then

$$f(p + q) = f(p) + f(q)$$

and

$$f(p \Rightarrow q) = (f(p) \Rightarrow f(q)).$$

It follows, in particular, that every Boolean homomorphism is a ring homomorphism, and also that every Boolean homomorphism is order preserving. The last assertion means that if $p \leq q$, then $f(p) \leq f(q)$.

The crucial fact in the preceding paragraph was the definability of Boolean operations in terms of meet, join, and complementation. Thus, more generally, if a mapping f from a Boolean algebra **B** to a Boolean algebra **A** preserves enough Boolean operations so that all others are definable in terms of them, then f is a homomorphism. Example: if f preserves $\vee$ and $\neg$ (that is, if f satisfies identities (ii) and (iii)), then f is a homomorphism; alternatively, if f preserves the Sheffer stroke, then f is a homomorphism.

There is a simple but important connection between homomorphisms and subalgebras. If f is a homomorphism from **B** to **A**, and if **C** is a subalgebra of **B**, then the *image of* **C** *under* f, that is, the set

$$f(\mathbf{C}) = \{f(p) : p \in \mathbf{C}\},$$

is easily seen to be a subalgebra of **A**. Similarly, if **D** is a subalgebra of **A**, then the *inverse image* of **D** *under* f, that is, the set

$$f^{-1}(\mathbf{D}) = \{p \in \mathbf{B} : f(p) \in \mathbf{D}\},$$

is a subalgebra of **B**.

28. Examples of homomorphisms

We proceed to consider some examples of Boolean homomorphisms. For our first example let **B** be an arbitrary Boolean algebra and let p_0 be an arbitrary element of **B**. The set **A** of all subelements of p_0 (that means elements p with $p \leq p_0$) can be construed as a Boolean algebra as follows: 0, meet, and join in **A** are the same as in **B**, but 1 and $\neg p$ in **A** are defined to be the elements p_0 and $p_0 - p$ of **B**. This Boolean algebra **A** is called the *relativization of* **B** *to* p_0; the mapping $p \mapsto p \wedge p_0$ is an **A**-valued homomorphism on **B**.

Consider next a field **B** of subsets of a set X, and let x_0 be an arbitrary point of X. For each set P in **B**, let $f(P)$ be 1 or 0 according as $x_0 \in P$ or $x_0 \notin P$. The mapping f is a **2**-valued homomorphism on **B**. Observe that $f(P)$ is equal to the value of the characteristic function of P at x_0. (Recall that the characteristic function of P is the function χ from X to 2 such that $\chi(p) = 1$ if p is in P, and $\chi(p) = 0$ otherwise.)

For one more example, let φ be an arbitrary mapping from a non-empty set X into a set Y, and let **A** and **B** be fields of subsets of X and Y respectively. Write $f = \varphi^{-1}$, or, explicitly, for each P in **B**, let $f(P)$ be the inverse image of P under φ. In general, the set $f(P)$ will not belong to the field **A**. If $f(P) \in \mathbf{A}$ whenever $P \in B$ (in particular, if **A** is the field of all subsets of X), then f is an **A**-valued homomorphism on **B**.

For purposes of reference, we shall call the homomorphisms described in these three examples the homomorphisms *induced* by p_0, x_0, and φ respectively.

If **B** is a subalgebra of an algebra **A**, then the identity mapping (that is, the mapping f defined for every p in **B** by $f(p) = p$) is a homomorphism from **B** into **A**, and, in particular, the identity mapping on **A** is an automorphism of **A**.

There is a natural way to define the product of (some) pairs of homomorphisms, and it turns out that, with respect to this product, the identity mappings just mentioned indeed act as multiplicative identities. The *product* (or composite)

$$g \circ h$$

of two homomorphisms is defined in case **A**, **B**, and **C** are Boolean algebras, g maps **B** into **C**, and h maps **A** into **B** (as in the figure below)

$$\mathbf{A} \underset{h}{\rightarrow} \mathbf{B} \underset{g}{\rightarrow} \mathbf{C}.$$

The value of $g \circ h$ at each element p of **A** is given by

$$(g \circ h)(p) = g(h(p)).$$

If, moreover, f is a homomorphism from **C** into, say, **D** (as in the figure below)

$$\mathbf{A} \underset{h}{\rightarrow} \mathbf{B} \underset{g}{\rightarrow} \mathbf{C} \underset{f}{\rightarrow} \mathbf{D},$$

then

$$f \circ (g \circ h) = (f \circ g) \circ h;$$

that is, the operation of composition is associative.

29. Free algebras

The elements of every subset of every Boolean algebra satisfy various algebraic conditions (such as the distributive law, for example) just because they belong to the same Boolean algebra. If the elements of some particular set **E** satisfy no conditions except these necessary universal ones, it is natural to describe **E** by some such word as "free". A crude but suggestive way to express the fact that the elements of **E** satisfy no special conditions is to say that the elements of **E** can be transferred to an arbitrary Boolean algebra in a completely arbitrary way with no danger of encountering a contradiction. In what follows we shall make these heuristic considerations precise. We shall restrict attention to sets that generate the entire algebra; from the practical point of view the loss of generality involved in doing so is negligible.

A set **E** of generators of a Boolean algebra **B** is called *free* if every mapping from **E** to an arbitrary Boolean algebra **A** can be extended to an **A**-valued homomorphism on **B**. In more detail: **E** is free in case, for every Boolean algebra **A** and for every mapping g from **E** into **A**, there exists an **A**-valued homomorphism f on **B** such that $f(p) = g(p)$ for every p in **E**. Equivalent expressions: "**E** freely generates **B**" or "**B** is free on **E**". A Boolean algebra is called free if it has a free set of generators.

The definition is conveniently summarized by the diagram.

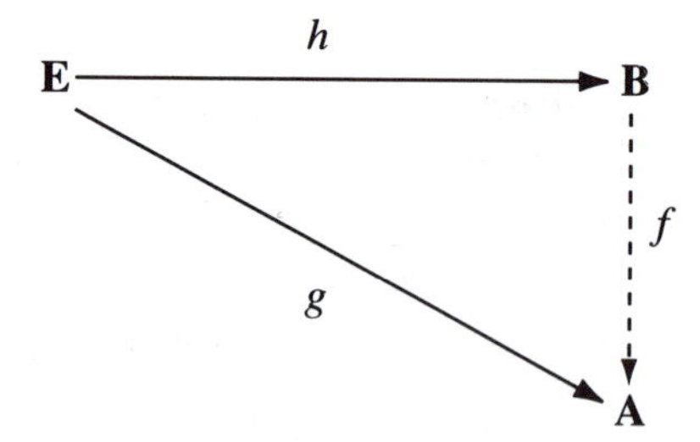

The diagram is to be interpreted as follows. The arrow h is the identity mapping from **E** to **B**, expressing the fact that **E** is a subset of **B**. The arrow g is an arbitrary mapping from **E** to an arbitrary algebra **A**. The arrow f is, of course, the homomorphic extension required by the definition; it is dotted to indicate that it comes last, as a construction based on h and g. It is understood that the diagram is "commutative" in the sense that

$$(f \circ h)(p) = g(p)$$

for every p in **E**.

The arrow diagram does not express the fact that **E** generates **B**. The most useful way that that fact affects the mappings under consideration is to guarantee uniqueness: there can be only one **A**-valued homomorphism f on **B** that agrees with g on $\mathbf{E}$. One way of expressing this latter fact is to say that f is uniquely determined by g and h.

There is another and even more important uniqueness assertion that can be made here. If $\mathbf{B}_1$ and $\mathbf{B}_2$ are Boolean algebras, free on subsets $\mathbf{E}_1$ and $\mathbf{E}_2$ respectively, and if $\mathbf{E}_1$ and $\mathbf{E}_2$ have the same cardinality, then $\mathbf{B}_1$ and $\mathbf{B}_2$ are isomorphic via an isomorphism that interchanges $\mathbf{E}_1$ and $\mathbf{E}_2$. The proof is summarized by the diagram below.

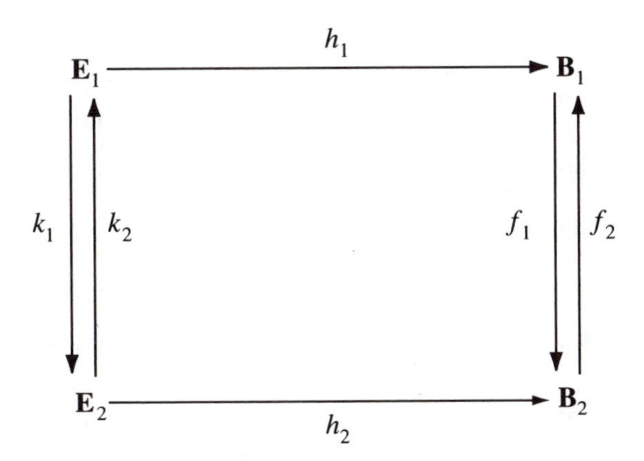

Here k_1 is a one-to-one mapping from $\mathbf{E}_1$ onto $\mathbf{E}_2$ with inverse k_2, the mappings h_1 and h_2 are the identity mappings on the sets $\mathbf{E}_1$ and $\mathbf{E}_2$, and f_1 and f_2 are the homomorphic extensions of k_1 and k_2 respectively (that is, $f_1 \circ h_1 = k_1$ and $f_2 \circ h_2 = k_2$). The commutativity of the diagram tells us that the two ways of going from $\mathbf{E}_1$ to $\mathbf{B}_2$ must coincide, and the same is true for the two ways of going from $\mathbf{E}_2$ to $\mathbf{B}_1$. If we apply the appropriate one of f_1 and f_2 to these equalities, and use the fact that the composite of k_1 and k_2, in either order, is the identity on its domain, we conclude that $f_2 \circ f_1$ and $f_1 \circ f_2$ are the extensions of h_1 and h_2 respectively. Since the identity homomorphisms on $\mathbf{B}_1$ and $\mathbf{B}_2$ are also such extensions, the already known uniqueness result guarantees that the composite of f_1 and f_2, in either order, is the identity on its domain. This implies that f_1 and f_2 are isomorphisms, and the proof is complete.

There is one big gap in what we have seen so far of the theory of freely generated algebras. We may know all about uniqueness, but we know nothing about existence. The main thing to be known here is that for each cardinal number there actually exists a Boolean algebra that is free on a set having exactly that many elements. In our development of Boolean algebra, motivated by the logical considerations of the propositional calculus, this result is *almost* available. In developing the propositional calculus we dealt with a prescribed countably infinite set, called the propositional variables. That development could have been carried out, with no significant change, with a quite arbitrary set playing the central role of the set of propositional variables — the cardinality of that set could have been finite, countable, or uncountable. The result is a Boolean algebra **A** with a certain set of generators (namely the equivalence classes corresponding to the propositional variables) that behave, at least in part, like free generators. They act free in this respect, that every mapping from them to the Boolean algebra **2** can be extended to a **2**-valued homomorphism. This, in fact, is what the discussion of "interpretations" was all about; an interpretation naturally induces (it is tempting to say that it is, but the propositional calculus is not the same as its Boolean algebra of equivalence classes of formulas) a homomorphism from **A** to **2**. The fact is that the Boolean algebras (one for each cardinal number) derived from the propositional calculus are indeed free, that, in fact, the (equivalence classes of) propositional variables constitute a free set of generators, but the proof of this fact depends on a representation theorem for Boolean algebras that will not be discussed till a little later.

We borrow the result from the future and assume known that "the Boolean algebra derived from the propositional calculus is free (with a countably infinite set of generators)." This is the point toward which the construction of the propositional calculus was aiming all along, and this point enables us to see what is meant by speaking of other "versions" of the propositional calculus. The propositional calculus may be approached via symbols completely different from the ones used above, via many radically different axiomatizations with many different sets of rules of procedure, or, for that matter, it needn't be approached via axioms and rules of inference at all. To say, nevertheless, that the same thing is being approached from all these different points of view means just that the end result of the construction is the free Boolean algebra with $\aleph_0$ generators.

30. Kernels and ideals

If f is a Boolean homomorphism, from **B** to **A**, say, the *kernel* of f is the set of those elements in **B** that f maps onto 0 in **A**. In symbols, the kernel **I** of f is defined by

$$\mathbf{I} = f^{-1}(\{0\}),$$

or, equivalently, by

$$\mathbf{I} = \{p : f(p) = 0\}.$$

Here is a general and useful remark about homomorphisms and their kernels: a necessary and sufficient condition that a homomorphism be a monomorphism (one-to-one) is that its kernel be trivial. Proof of necessity: if f is one-to-one and $f(p) = 0$, then $f(p) = f(0)$, and therefore $p = 0$. Proof of sufficiency: if the kernel of f is $\{0\}$ and if $f(p) = f(q)$, then

$$f(p+q) = f(p) + f(q) = 0,$$

so that $p + q = 0$, and this means that $p = q$. (Recall that, as a ring, a Boolean algebra has characteristic 2; in other words, every element is its own additive inverse.)

Motivated by immediately obvious properties of kernels, we make the following definition: a *Boolean ideal* in a Boolean algebra **B** is a subset **I** of B such that

(i) $$0 \in \mathbf{I},$$

(ii) $$\text{if} \quad p \in \mathbf{I} \quad \text{and} \quad q \in \mathbf{I}, \quad \text{then} \quad p \vee q \in \mathbf{I},$$

(iii) $$\text{if} \quad p \in \mathbf{I} \quad \text{and} \quad q \in \mathbf{B}, \quad \text{then} \quad p \wedge q \in \mathbf{I}.$$

Observe that condition (i) in the definition can be replaced by the superficially less restrictive condition that **I** be not empty, without changing the concept of ideal. Indeed, if **I** is not empty, say $p \in \mathbf{I}$, and if **I** satisfies (iii), then $p \wedge 0 \in \mathbf{I}$. The condition (iii) itself can be replaced by

$$\text{if} \quad p \in \mathbf{I} \quad \text{and} \quad q \leq p, \quad \text{then} \quad q \in \mathbf{I}$$

without changing the concept of ideal; the proof is elementary.

Clearly the kernel of every Boolean homomorphism is a Boolean ideal. Thus, every example of a homomorphism gives rise to an example of an ideal: its kernel. For instance, if f is the homomorphism on **B** induced by (relativization to) p_0, namely $f(p) = p \wedge p_0$ for every p, then the corresponding ideal consists of all those elements p for which $p \wedge p_0 = 0$, or, equivalently, $p \leq \neg p_0$. If f is defined on a field of subsets of X so that $f(P)$ is the value of the characteristic function of P at some particular point x_0 of X, then the corresponding ideal consists of all those sets P in the field that do not contain x_0. If, finally, the homomorphism f is induced by a mapping φ from a set X into a set Y, then the corresponding ideal consists of all those sets P in the domain of f that are disjoint from the range of φ.

There are examples of ideals for which it is not obvious that they are associated with some homomorphism. One such example is the class of all finite subsets in the field of all subsets of a set. More generally, the collection of all those finite sets that happen to belong to some particular field (not necessarily the field of all subsets) is an ideal in that field.

Every Boolean algebra **B** has a *trivial* ideal, namely the set $\{0\}$ consisting of 0 alone; all other ideals of **B** will be called *non-trivial.* Every Boolean algebra **B** has an *improper* ideal, namely **B** itself; all other ideals will be called *proper.* Observe that an ideal is proper if and only if it does not contain 1. therefore an ideal is improper if and only if it contains both p and $\neg p$, for some (or, equivalently, for all) p in **B**.

The intersection of every collection of ideals in a Boolean algebra **B** is again an ideal of **B**. It follows that if **E** is an arbitrary subset of **B**, then the intersection of all those ideals of **B** that happen to include **E** is an ideal. (There is always at least one ideal that includes **E**, namely the improper ideal **B**.) That intersection, say **I**, is the smallest ideal that includes **E**; in other words, **I** is included in every ideal in **B** that includes **E**. The ideal **I** is called the ideal *generated* by **E**. Thus, for example, if **E** is empty, then the ideal generated by **E** is the smallest possible ideal of **B**, namely the trivial ideal $\{0\}$. An ideal generated by a (not empty) singleton $\{p\}$ is called a *principal* ideal; it consists of all the elements that are less than or equal to p.

As in the case of the subalgebra generated by a (not empty) set **E**, one can give a "bottom-up" construction of the ideal **I** generated by **E**. For any subset X of **E**, write

$$X^* = \{p \vee q : p, q \in X\} \cup \{p \wedge q : p \in X \text{ and } q \in \mathbf{B}\}.$$

If $I_0 = \mathbf{E}$ and

$$I_{n+1} = I_n^* \cup I_n,$$

then it is not difficult to check that the union of the sets I_n is just the ideal **I** generated by **E**.

The above construction clarifies the sense in which **E** really does "generate" **I**. Here is another characterization that is much more practical: an element q of **B** is in the ideal **I** generated by a set **E** if and only if there is a finite sequence $p_1, \ldots, p_n$ of elements in **E** such that

$$q \leq p_1 \vee \cdots \vee p_n.$$

Indeed, let $\mathbf{J}$ be the set of all those q for which such a sequence exists. It is immediate from the definition of an ideal, applied to $\mathbf{I}$, that the join of elements from $\mathbf{E}$, and hence each element below such a join, must be in $\mathbf{I}$. Thus $\mathbf{J}$ is included in $\mathbf{I}$. On the other hand, it is simple to verify that $\mathbf{J}$ is, in fact, an ideal and includes $\mathbf{E}$. Since $\mathbf{I}$ is the smallest ideal that includes $\mathbf{E}$, it follows that $\mathbf{I}$ is included in $\mathbf{J}$.

There is an important special case of this characterization that is often of use: if $\mathbf{I}$ is an ideal of $\mathbf{B}$ and $p_0 \in \mathbf{B}$, then the ideal generated by $\mathbf{I} \cup \{p_0\}$ is just the set

$$\mathbf{J} = \{p \vee q : p \leq p_0 \text{ and } q \in \mathbf{I}\}.$$

The proof is an easy exercise.

If $\mathbf{I}$ is a proper ideal, then, for each element p_0, the ideal generated by $\mathbf{I} \cup \{p_0\}$ is proper just in case $\neg p_0 \notin \mathbf{I}$. To prove this, let $\mathbf{J}$ be the ideal generated by $\mathbf{I} \cup \{p_0\}$. Since it contains p_0, the ideal $\mathbf{J}$ is proper just in case it does not contain $\neg p_0$. It remains to show that $\neg p_0$ is in $\mathbf{J}$ if and only if it is in $\mathbf{I}$. If $\neg p_0$ is in $\mathbf{I}$, then obviously it is in $\mathbf{J}$. Now suppose that $\neg p_0 \in \mathbf{J}$. Then

$$\neg p_0 = p \vee q$$

for some $p \leq p_0$ and some $q \in \mathbf{I}$. Taking the meet of both sides of this equation with $\neg p_0$, we obtain

$$\neg p_0 = \neg p_0 \wedge q.$$

Since $\neg p_0 \wedge q$ is in $\mathbf{I}$ (because q is in $\mathbf{I}$), we conclude that $\neg p_0$ is in $\mathbf{I}$.

The union of a non-empty collection $\mathcal{C}$ of ideals is not necessarily an ideal. If, however, $\mathcal{C}$ is linearly ordered by inclusion (that is, whenever two ideals are in $\mathcal{C}$, one of them includes the other), then the union is an ideal. For example, to verify that the union is closed under the formation of joins, assume that p and q are in the union. Then there are ideals $\mathbf{I}$ and $\mathbf{J}$ in $\mathcal{C}$ such that $p \in \mathbf{I}$ and $q \in \mathbf{J}$. Because $\mathcal{C}$ is linearly ordered, one of these ideals is included in the other, say $\mathbf{I} \subset \mathbf{J}$. Since both p and q are in $\mathbf{J}$, their join $p \vee q$ is in $\mathbf{J}$, and hence also in the union. The verification of the other conditions is a simple exercise. Notice that the union of the ideals in $\mathcal{C}$ is a proper ideal just in case each of the ideals in $\mathcal{C}$ is proper (because 1 is in the union just in case it is in one of the ideals).

The concepts of subalgebra and homomorphism are in a certain obvious sense self-dual: if we interchange the meet and join operations everywhere in the definitions of these notions, we obtain the same notions. The concept of ideal is not self-dual. The dual concept is defined as follows. A *Boolean filter* in a Boolean algebra **B** is a subset **F** of B such that

(iv) $$1 \in \mathbf{F},$$

(v) $$\text{if} \quad p \in \mathbf{F} \quad \text{and} \quad q \in \mathbf{F}, \quad \text{then} \quad p \wedge q \in \mathbf{F},$$

(vi) $$\text{if} \quad p \in \mathbf{F} \quad \text{and} \quad q \in \mathbf{B}, \quad \text{then} \quad p \vee q \in \mathbf{F}.$$

The condition (iv) can be replaced by the condition that **F** be not empty. The condition (vi) can be replaced by

$$\text{if} \quad p \in \mathbf{F} \quad \text{and} \quad p \leq q, \quad \text{then} \quad q \in \mathbf{F}.$$

Neither of these replacements will alter the concept being defined. The filter *generated* by a subset of **B** and, in particular, a *principal* filter, are defined by an obvious dualization of the corresponding definitions for ideals. Here is the corresponding (dual) characterization of the notion of the filter generated by a set **E**: an element q is in the filter **F** generated by **E** if and only if there is a finite sequence $p_1, \ldots, p_n$ of elements in **E** such that

$$p_1 \wedge \cdots \wedge p_n \leq q.$$

The relation between filters and ideals is a very close one. The fact is that filters and ideals come in dual pairs. This means that there is a one-to-one correspondence that pairs each ideal to a filter, its dual, and by means of which every statement about ideals is immediately translatable to a statement about filters. The pairing is easy to describe. If **I** is an ideal, write

$$\mathbf{F} = \{p : \neg p \in \mathbf{I}\},$$

and, in reverse, if **F** is a filter, write

$$\mathbf{I} = \{p : \neg p \in \mathbf{F}\}.$$

It is trivial to verify that this construction does indeed convert an ideal into a filter, and vice versa.

A congruence relation on a Boolean algebra **B** is an equivalence relation $\equiv$ that is compatible with the operations of **B** in the following sense. If p, q, r, and s are elements of **B**, and if $p \equiv r$ and $q \equiv s$, then

$$\neg p \equiv \neg r,$$

$$p \vee q \equiv r \vee s,$$

and

$$p \wedge q \equiv r \wedge s.$$

If $\equiv$ is a congruence relation in **B**, then the set $\{p : p \equiv 0\}$ of elements congruent to 0 forms an ideal, called the *kernel* of the relation. The corresponding filter is just the dual set $\{p : p \equiv 1\}$ of elements congruent to 1. Conversely, given any ideal **I** of **B**, the relation $\equiv$ defined by

$$p \equiv q \qquad \text{if and only if} \qquad p + q \in \mathbf{I}$$

is a congruence relation on **B**; it is called the congruence relation *induced* by **I**. Finally, the congruence induced by the kernel of a congruence relation $\equiv$ is just $\equiv$ again, and the kernel of the congruence induced by an ideal **I** is just **I** again. Thus, every congruence uniquely determines, and is uniquely determined by, an ideal, namely its kernel. It is of course necessary to check all these assertions, but there is no difficulty about that. Ideals, filters, and congruence relations are just three different ways of talking about the same thing.

Every congruence relation in an algebra induces a corresponding quotient algebra. Here is how. Suppose that $\equiv$ is a congruence relation on **B** with kernel **I**. Let $\mathbf{B}/\mathbf{I}$ be the set of of all equivalence classes of $\equiv$. (These equivalence classes are called the *cosets* of **I**). We can turn $\mathbf{B}/\mathbf{I}$ into a Boolean algebra by defining the Boolean operations as follows: if P is an equivalence class, with $p \in P$, let $\neg P$ be the equivalence class to which $\neg p$ belongs; if P and Q are equivalence classes with $p \in P$ and $q \in Q$, let $P \vee Q$ and $P \wedge Q$ be the equivalence classes to which $p \vee q$ and $p \wedge q$ belong. A routine verification shows that these definitions are unambiguous and that what they define is a Boolean algebra. This algebra is called the *quotient of* **B** *modulo* **I**, and is also denoted by $\mathbf{B}/\mathbf{I}$. By the quotient of a Boolean algebra modulo an ideal or a filter, we mean the quotient modulo the induced congruence relation.

Note that the concept of a quotient algebra already occurred (implicitly) above; it was used to make a Boolean algebra out of the propositional calculus.

31. Maximal ideals

An ideal is *maximal* if it is a proper ideal that is not properly included in any other proper ideal. Equivalently, to say that **I** is a maximal ideal in **B** means that **I** is a proper ideal and **B** is the only ideal that properly includes **I**. Examples:

the trivial ideal is maximal in **2**; the ideals, in fields of sets, defined by the exclusion of one point are maximal.

Maximal ideals are characterized by a curious algebraic property: an ideal **I** in a Boolean algebra **B** is maximal if and only if, for each p in **B**, either $p \in \mathbf{I}$ or $\neg p \in \mathbf{I}$, but not both.

To prove this, assume first that, for some p in **B**, neither p nor $\neg p$ is in **I**. If **J** is the ideal generated by $\mathbf{I} \cup \{p\}$, then $\mathbf{J} \neq \mathbf{I}$, since $p \notin \mathbf{I}$; also, **J** is proper since $\neg p \notin \mathbf{J}$. Therefore, **I** is not maximal. To prove the converse, assume that always either p or $\neg p$ is in **I**, and suppose that **J** is an ideal properly including **I**; it is to be proved that **J** is improper. Since $\mathbf{J} \neq \mathbf{I}$, there is an element p in **J** that does not belong to **I**. The assumption implies that $\neg p \in \mathbf{I}$, and therefore $\neg p \in \mathbf{J}$; the assertion is proved.

Here is a closely related characterization of maximal ideals: an ideal **I** in a Boolean algebra **B** is maximal if and only if the quotient algebra $\mathbf{B}/\mathbf{I}$ is (isomorphic to) **2**.

One direction of the proof is almost immediate. Assume that $\mathbf{B}/\mathbf{I}$ is **2**. Then **I** must be a proper ideal, for otherwise the quotient would be trivial. Let $[p]$ denote the equivalence class of p in $\mathbf{B}/\mathbf{I}$. If $p \notin \mathbf{I}$, then $[p] = 1$ and therefore $[\neg p] = \neg [p] = 0$, whence $\neg p \in \mathbf{I}$. This shows that **I** is maximal.

Now assume that **I** is maximal. Since **I** is proper, the quotient $\mathbf{B}/\mathbf{I}$ has at least two elements. To see that it has at most two elements, let p and q be in **B** but not in **I**. The assumption of maximality implies that $\neg p$ and $\neg q$ are in **I**. Therefore the elements $\neg p \wedge q$ and $p \wedge \neg q$, and hence also

$$p + q = (\neg p \wedge q) \vee (p \wedge \neg q),$$

are in **I**. Conclusion: $[p] = [q]$.

Does every non-trivial Boolean algebra have maximal ideals? The answer is yes. In fact, more is true.

Maximal ideal theorem. *Every proper ideal in a Boolean algebra is included in some maximal ideal.*

We shall prove only the special case of the theorem when the algebra in question is countable. Let **B** be a countable Boolean algebra, **I** a proper ideal in **B**, and $p_0, p_1, p_2, \ldots$ an enumeration of the elements in **B**. We define inductively a sequence $\mathbf{J}_0, \mathbf{J}_1, \mathbf{J}_2, \ldots$ of proper ideals as follows. Take $\mathbf{J}_0$ to be **I**. Assume, now, that $\mathbf{J}_n$ has been defined. If $\neg p_n$ is in $\mathbf{J}_n$, then take $\mathbf{J}_{n+1}$ to be $\mathbf{J}_n$.

Otherwise, take $\mathbf{J}_{n+1}$ to be the ideal generated by $\mathbf{J}_n \cup \{p_n\}$. A straightforward argument by induction on n, using the observations of the preceding section, shows that each ideal $\mathbf{J}_n$ is proper and is included in the ideal $\mathbf{J}_{n+1}$. Therefore the union $\mathbf{M}$ of these ideals is a proper ideal and clearly includes $\mathbf{I}$. To show that $\mathbf{M}$ is maximal, we must show that for each natural number n, either p_n or $\neg\, p_n$ is in $\mathbf{M}$. If $\neg\, p_n$ is in $\mathbf{J}_n$, then it is obviously in $\mathbf{M}$. If $\neg\, p_n$ is not in $\mathbf{J}_n$, then we put p_n into $\mathbf{J}_{n+1}$ and hence into $\mathbf{M}$. It follows from the characterization of maximality that $\mathbf{M}$ is maximal.

To prove the theorem in the case when $\mathbf{B}$ is not countable, we need to know that the elements of $\mathbf{B}$ can be enumerated in a transfinite sequence indexed by ordinals. The existence of such an enumeration requires one of the basic axioms of set theory, the so-called "axiom of choice". Once such an enumeration is in hand, however, the proof that $\mathbf{I}$ can be extended to a maximal ideal is nearly identical to the proof in the countable case, except that the construction of the sequence of ideals takes transfinitely many steps.

We shall omit the details of this transfinite construction. The remarks of the last two paragraphs should give readers a good intuition of what is involved. Before we apply the maximal ideal theorem, one example and one word of warning are in order.

The assertion of the theorem is that if $\mathbf{I}$ is a proper ideal in a Boolean algebra $\mathbf{B}$, then there exists at least one maximal ideal that includes $\mathbf{I}$; in general, there are many such maximal ideals. Suppose, for instance, that $\mathbf{B}$ is the field of all subsets of a set X, and suppose that P is a non-empty subset of X. If $\mathbf{I}$ is the collection of all subsets of $X - P$, then $\mathbf{I}$ is a proper ideal. If $x \in P$, and if $\mathbf{M}_x$ is the collection of all subsets of $X - \{x\}$, then $\mathbf{M}_x$ is a maximal ideal that includes $\mathbf{I}$, and this is true for each x in P.

That is the example; the word of warning is that not every example is so transparent. The maximal ideal theorem is an existential assertion that is highly non-constructive; in most cases no inspection, however thorough, will reveal what a maximal ideal of the kind it talks about looks like. Suppose, for instance, that $\mathbf{B}$ is the field of all subsets of an infinite set X, and let $\mathbf{I}$ be the collection of all finite subsets of X. The maximal ideal theorem guarantees the existence of a maximal ideal that contains all finite sets, but no such maximal ideal is visible. One thing is sure: no such maximal ideal can have the form $\mathbf{M}_x$ (or, equivalently, no such maximal ideal can be principal).

32. Homomorphism theorem

The definition of ideals was formulated so as to guarantee that the kernel of every homomorphism is an ideal. It is natural and important to raise the converse question: is every ideal the kernel of some homomorphism? For maximal ideals the answer is easily seen to be yes. Suppose, indeed, that $\mathbf{I}$ is a maximal ideal in $\mathbf{B}$, and write $f(p) = 0$ or 1, according as the element p belongs to $\mathbf{I}$ or not. In view of the observation in the section on ideals, the definition of f can also be formulated this way: $f(p) = 0$ or 1, according as $p \in \mathbf{I}$ or $\neg p \in \mathbf{I}$. A straightforward verification, based on the characterization of maximal ideals at the beginning of the preceding section, shows that f is a homomorphism from $\mathbf{B}$ to $\mathbf{2}$; the kernel of f is obviously $\mathbf{I}$. What we have proved in this way is a very special case of the following result.

The homomorphism theorem. *Every ideal is the kernel of an epimorphism.*

A similar result is standard in group theory, ring theory, and many other parts of algebra; it depends on the universal algebraic concept of quotient structure that was discussed above. To prove the theorem, let $\mathbf{B}$ be a Boolean algebra and $\mathbf{I}$ an ideal in $\mathbf{B}$. The desired epimorphism is the so-called natural or *canonical* homomorphism, or projection, from $\mathbf{B}$ onto the quotient $\mathbf{B}/\mathbf{I}$; it associates with each element of $\mathbf{B}$ the equivalence class (or coset) of that element modulo $\mathbf{I}$.

33. Consequences

Associated with the homomorphism theorem there is a cluster of results of the universal algebraic kind, some of which we now proceed to state.

Suppose that $\mathbf{I}$ is an ideal in a Boolean algebra $\mathbf{B}$, write $\mathbf{A} = \mathbf{B}/\mathbf{I}$, and let f be the canonical homomorphism from $\mathbf{B}$ onto $\mathbf{A}$. The mapping that associates with every ideal $\mathbf{J}$ in $\mathbf{A}$ the set $f^{-1}(\mathbf{J})$ in $\mathbf{B}$ is a one-to-one correspondence between all the ideals in $\mathbf{A}$ and all those ideals in $\mathbf{B}$ that include $\mathbf{I}$. The images of the trivial ideal and of $\mathbf{A}$ under this correspondence are $\mathbf{I}$ and $\mathbf{B}$ respectively. If $\mathbf{J}_1 \subset \mathbf{J}_2$, then $f^{-1}(\mathbf{J}_1) \subset f^{-1}(\mathbf{J}_2)$. If f_0 is a homomorphism from $\mathbf{B}$ to a Boolean algebra $\mathbf{A}_0$, say, and if the kernel $\mathbf{I}_0$ of f_0 includes $\mathbf{I}$, then there exists a unique homomorphism g from $\mathbf{A}$ to $\mathbf{A}_0$ such that $f_0 = g \circ f$.

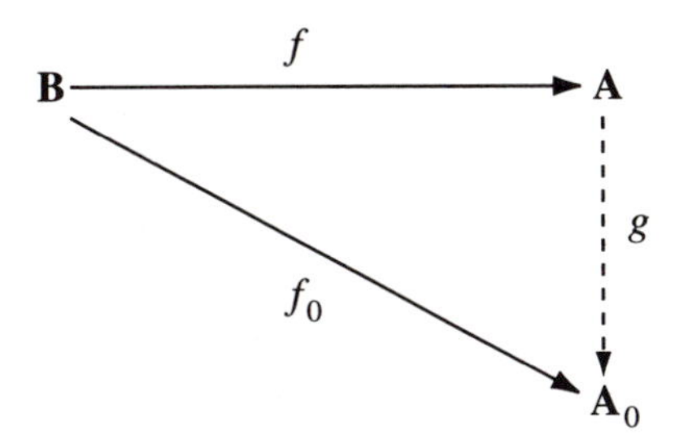

The proofs of all these assertions are the same for Boolean algebras as for other algebraic structures (such as groups and rings); the words may change, but the ideas stay the same. It is not worthwhile to record the proofs here; the interested reader should have no difficulty in reconstructing them.

A Boolean algebra is called *simple* if it is not trivial and has no non-trivial proper ideals. Simplicity is a universal algebraic concept, but, as it turns out, in the context of Boolean algebras it is not a fruitful one. The reason is that, up to isomorphic copies, there is exactly one simple algebra, namely **2**. Clearly **2** is simple. If, conversely, **B** is simple, and if p is a non-zero element of **B**, then the principal ideal generated by p must be improper, which can happen only if $p = 1$. In other words, **B** is such that if an element of **B** is not 0, then it is 1; this means that $\mathbf{B} = \mathbf{2}$.

This concludes our study of the universal aspects of Boolean algebra; we turn now to its more provincial aspects.

34. Representation theorem

If a Boolean algebra **A** is a field of subsets of a set X, then the points of X serve to define **2**-valued homomorphisms of **A**. (Recall that each x_0 of X induces the homomorphism that sends a set P in **A** to 1 or 0, according as $x_0 \in P$ or $x_0 \notin P$.) This comment suggests that if we start with a Boolean algebra **A** and seek to represent it as a field of subsets of some set X, a reasonable place to conduct the search for points suitable to make up X is among the **2**-valued homomorphisms of **A**. The suggestion would be impractical if it turned out that **A** has no **2**-valued homomorphisms. Our first result along these lines is that there is nothing to fear; there is always a plethora of **2**-valued homomorphisms.

Existence theorem. *For every non-zero element p_0 of every Boolean algebra* **A**, *there is a* **2**-*valued homomorphism x on* **A** *such that* $x(p_0) = 1$.

In view of the comments that precede the homomorphism theorem, the conclusion of the existence theorem can be rephrased as follows: there exists a maximal ideal **M** in **A** such that $p_0 \notin \mathbf{M}$. To get one, apply the maximal ideal theorem to the ideal **I** of all those elements p of **A** for which $p \leq \neg p_0$. Clearly $p_0 \notin \mathbf{M}$, for otherwise the element $1 = p_0 \vee \neg p_0$ would be in **M**.

The existence theorem is actually equivalent to the maximal ideal theorem in the sense that each is easily derivable from the other. Indeed, the proof of the existence theorem shows that it follows from the maximal ideal theorem. To obtain the reverse implication, suppose that **I** is a proper ideal in **B**. Then there is an element p in **B** that is not in **I**. In the quotient algebra $\mathbf{B}/\mathbf{I}$ the element $[p]$ (the equivalence class of p) is non-zero. By the existence theorem, there is a **2**-valued homomorphism f on $\mathbf{B}/\mathbf{I}$ such that $f([p]) \neq 0$. Let g be the canonical homomorphism of **B** onto $\mathbf{B}/\mathbf{I}$. Then the composition $f \circ g$ is a **2**-valued homomophism on **B**. Therefore its kernel **M** is a proper ideal. Because $f \circ g$ maps each element of **I** to 0, the ideal **M** includes **I**. Because $f \circ g$ is **2**-valued, the quotient **B/M** is (isomorphic to) **2**. Therefore **M** must be maximal.

We are now ready for the proof of the principal result.

Representation theorem (M. H. Stone). *Every Boolean algebra is isomorphic to a field of sets.*

Proof. Given a Boolean algebra **A**, let X be the set of all **2**-valued homomorphisms on **A**, and, for each p in **A**, write

$$f(p) = \{x \in X : x(p) = 1\}.$$

The mapping f associates a subset of X with each element of **A**. The verification that f is a Boolean homomorphism from **A** onto a subalgebra of the field of subsets of X is purely mechanical. Thus

$$\begin{aligned} f(p \vee q) &= \{x : x(p \vee q) = 1\} \\ &= \{x : x(p) \vee x(q) = 1\} \\ &= \{x : x(p) = 1\} \cup \{x : x(q) = 1\} \\ &= f(p) \cup f(q) \end{aligned}$$

and

$$\begin{aligned} f(\neg p) &= \{x : x(\neg p) = 1\} \\ &= \{x : \neg x(p) = 1\} \end{aligned}$$

$$
\begin{aligned}
&= \{x : x(p) \neq 1\} \\
&= X - \{x : x(p) = 1\} \\
&= X - f(p).
\end{aligned}
$$

If $f(p) = 0$, that is, if $\{x : x(p) = 1\} = \varnothing$, then $p = 0$ (by the existence theorem); this means that f is one-to-one. The proof is complete.

The preceding argument shows that the representation theorem is a simple consequence of the existence theorem. The converse is also true: the existence theorem follows readily from the representation theorem. For a proof, let **B** be a Boolean algebra and P a non-zero element of **B**. By the representation theorem, we may assume that **B** is a field of subsets of some set. The assumption that P is non-zero just means that P is not empty. Fix x_0 in P. For each set Q in **B**, let $f(Q)$ be 1 or 0 according as $x_0 \in Q$ or $x_0 \notin Q$. Then f is a **2**-valued homomorphism on **B**, and $f(P) = 1$.

What has been achieved so far in this section? We have proved the existence theoremand the representation theorem, and we have shown that each of them, as well as the maximal ideal theorem, is readily obtained from any one of the others.

As the first application of the representation theorem, let us prove that the propositional calculus does indeed induce a free Boolean algebra. Specifically, we shall prove that if **E** is a set of generators of a Boolean algebra **B** such that every mapping from **E** to **2** has an extension to a **2**-valued homomorphism on **B**, then **B** is free on **E**—that is, then every mapping from **B** to an arbitrary Boolean algebra **A** has an extension to an **A**-valued homomorphism on **B**.

Suppose indeed that g is a mapping from **E** to **A**. We must construct a homomorphism f from **B** into **A** that extends g. In view of the representation theorem, there is no loss of generality in assuming that **A** is a subalgebra (subfield) of the field **P** of all subsets of a set X, and we do so assume. If, in particular, $p \in$ **E**, then $g(p)$ is a subset of X. Denote the characteristic function of a subset Y of X by χ_Y. For each x in X the correspondence

$$p \mapsto \chi_{g(p)}(x)$$

maps **E** to **2**, and, consequently, has an extension f_x to a **2**-valued homomorphism on **B**. For each p in **B**, write

$$f(p) = \{x \in X : f_x(p) = 1\}.$$

Clearly $f(p)$ is a subset of X. It is easy to verify that f is in fact a homomorphism; the argument is similar to the one given in the proof of the representation theorem. If $p \in \mathbf{E}$, then

$$f_x(p) = \chi_{g(p)}(x),$$

and therefore

$$f(p) = \{x \in X : \chi_{g(p)}(x) = 1\} = g(p);$$

the homomorphism f from $\mathbf{B}$ to $\mathbf{P}$ agrees with g on $\mathbf{E}$. This says, among other things, that f maps $\mathbf{E}$ into $\mathbf{A}$. Since $f^{-1}(\mathbf{A})$ (that is, the set of all those p in $\mathbf{B}$ for which $f(p) \in \mathbf{A}$) is a Boolean subalgebra of $\mathbf{B}$, and since, as we have just seen, $\mathbf{E} \subset f^{-1}(\mathbf{A})$, it follows that $\mathbf{B} = f^{-1}(\mathbf{A})$. (Recall that $\mathbf{E}$ generates $\mathbf{B}$; there can be no proper subalgebra of $\mathbf{B}$ including $\mathbf{E}$.) This implies that the homomorphism f, already known to agree with g on $\mathbf{E}$, maps $\mathbf{B}$ into $\mathbf{A}$; the proof is complete.

Logic via algebra

35. Pre-Boolean algebras

Let us now cast a glance backward. The propositional calculus, as we first defined it, consisted of a set of "sentences". By definition, a sentence is a sequence satisfying certain conditions, whose terms belong to a certain "alphabet". The propositional calculus resembles a Boolean algebra in at least this respect: it is a set with three distinguished operations, one unary ($\neg$) and two binary ($\vee$ and $\wedge$). The operations are algebraically peculiar in that they are subject to no conditions whatever. Let us, tentatively and temporarily, refer to a structure like this (that is, like the propositional calculus) as a pre-Boolean algebra.

The skeleton of the propositional calculus acquires a little meat when theorems are defined. Regardless of how they are defined, the end results are a subset of the pre-Boolean algebra and an equivalence relation between the elements of the pre-Boolean algebra that behave like a filter and its induced congruence relation.

Let us make these parables precise and official. The way the propositional calculus was actually defined above and the way Boolean algebras were defined are different in a conceptually trivial respect: two propositional connectives ($\neg$ and $\vee$) were used in the one case and three operations ($\neg$, $\vee$, and $\wedge$) in the other. This was mere lack of foresight: there would have been nothing wrong with introducing all three connectives right away, or in restricting attention to only two operations. We ask the reader to make a suitable psychological adjustment (to, say, three connectives, corresponding to the three Boolean operations) needed to establish the parallelism between the propositional calculus and Boolean algebras.

There is another trivial, but possibly confusing, difference between the approach via logic and the approach via algebra. In the logic approach the notion of the provability of a sentence x, rather than the dual notion of the provability of its negation, $\neg x$, played the crucial role. The set of provable sentences has properties very similar to those of a filter, while the set of sentences whose negation is provable has properties similar to those of an ideal. In the algebra approach, ideals, rather than their duals, filters, were emphasized in defining the concept of quotient structure. This, too, is more of historical and emotional significance than conceptual, and we ask the reader, again, to adjust by adopting the algebraically uncomfortable but logically more desirable attitude of using filters. (The quotient of a Boolean algebra by a filter is the same as the quotient by the corresponding ideal.)

With this preparation, we are now in a position to make some precise definitions. (Warning: the concepts about to be defined are not standard parts of the subject.) We shall say, as adumbrated above, that a *pre-Boolean algebra*, or a B-algebra for short, is a set **B**, together with three operations, one unary ($\neg$) and two binary ($\vee$ and $\wedge$). Example: the set of sentences of the propositional calculus, in Polish (parenthesis-free) notation, with the operations of negation, disjunction, and conjunction. (The example could just as well have been given in terms of the classical notation, but then, as always, a special effort would have to be made about parentheses.) Another example: an arbitrary Boolean algebra, with complementation ($\neg$), join ($\vee$), and meet ($\wedge$).

Artificial examples could be added by the dozen. Consider, for instance, the positive integers with the successor operation ($n \mapsto n + 1$), addition, and multiplication, or the real numbers with the null operations

$$\neg x = 0, \quad x \vee y = 0, \quad \text{and} \quad x \wedge y = 0$$

for all x and y.

A *Boolean congruence* in a B-algebra **B** is an equivalence relation **F** that is a congruence relation for the distinguished operations ($\neg$, $\vee$, $\wedge$) and that converts **B** into a Boolean algebra. We use here a symbol such as **F** to denote the congruence relation because we have in mind the kind of relation induced by a filter in a Boolean algebra. In case **B** *is* a Boolean algebra and **F** *is* a filter, we shall identify **F** and the relation it induces. If p and q are in **B**, write

$$p \equiv q\,(\mathbf{F})$$

(or, perhaps, when it doesn't lead to ambiguity, just $p \equiv q$) to denote that p and q are equivalent in the sense of **F** (that is, p and q belong to the same **F** equivalence class). To say that **F** is a congruence relation means that if $p_1 \equiv p_2$

and $q_1 \equiv q_2$, then

$$\neg p_1 \equiv \neg p_2,$$
$$p_1 \vee q_1 \equiv p_2 \vee q_2,$$
$$p_1 \wedge q_1 \equiv p_2 \wedge q_2.$$

To say that **F** converts **B** into a Boolean algebra means that if "=" is replaced by "≡" in the axioms of a Boolean algebra, then the elements and operations of **B** satisfy those axioms. Equivalently, to say that **F** converts **B** into a Boolean algebra means that the equivalence classes of **B** modulo **F** form a Boolean algebra. (The operations on equivalence classes are defined as usual. If, for each p in **B**, the equivalence class of p is $[p]$, then

$$\neg [p] = [\neg p],$$
$$[p] \vee [q] = [p \vee q],$$
$$[p] \wedge [q] = [p \wedge q].$$

The unambiguity of these definitions follows from the fact that **F** is a congruence.)

We shall denote the Boolean algebra thus associated with a B-algebra **B** and a Boolean congruence **F** in **B** by **B/F** and call it the *quotient algebra*, **B** modulo **F**. The unit of this quotient algebra, that is, the set of elements of **B** that are equivalent to some element of the form $q \vee \neg q$, satisfies the three conditions that define a filter, and is therefore usually referred to as a filter (even though **B** may not be a Boolean algebra). Just as in the case of Boolean algebras, the filter induced by a Boolean congruence in a B-algebra uniquely determines the congruence. Unlike the situation for Boolean algebras, however, not every filter in a B-algebra uniquely determines a congruence, much less a Boolean congruence.

In harmony with the notational convention that identifies a Boolean filter with the congruence relation it induces, if **F** is a Boolean congruence in a B-algebra (not necessarily a Boolean algebra), we shall denote by the same symbol **F** the corresponding filter, that is, the unit element of the quotient algebra **B/F**. In other words, **F** may be viewed as a subset (namely an equivalence class) of **B**. A statement such as $p \in \mathbf{F}$ makes sense: it means that $[p] = 1$, that is, the equivalence class to which p belongs is the unit element of the quotient algebra **B/F**.

Here is an example of a Boolean congruence: in the B-algebra of all sentences of the propositional calculus, let **F** be the relation according to which

$$p \equiv q$$

means that the sentence

$$(\neg p \vee q) \wedge (p \vee \neg q)$$

is a theorem. Since this is the same as saying that

$$p \Leftrightarrow q$$

is a theorem, we shall refer to **F** as the *relation of provable equivalence*. Another example (mentioned several times above): any Boolean filter in any Boolean algebra induces the congruence defined by $p \equiv q$ if and only if the element $p \Leftrightarrow q$ is in the filter.

36. Substitution rule

Associated with every algebraic structure, even one with as little required structure as that of a B-algebra, there is the concept of homomorphism, defined as a structure-preserving mapping. Precisely: if $\mathbf{B}_1$ and $\mathbf{B}_2$ are B-algebras, a *homomorphism* from $\mathbf{B}_1$ to $\mathbf{B}_2$ is a mapping f from $\mathbf{B}_1$ to $\mathbf{B}_2$ such that

$$f(\neg p) = \neg f(p),$$

$$f(p \vee q) = f(p) \vee f(q),$$

and

$$f(p \wedge q) = f(p) \wedge f(q)$$

for all p and q in $\mathbf{B}_1$. The related concepts (such as endomorphism) are then defined in terms of homomorphisms as usual.

In terms of B-algebras and their homomorphisms, the complicated and treacherous substitution rule can be illuminated. The idea of that rule is that the propositional variables of the propositional calculus really are variables, and, as such, may be replaced by anything. To put this into exact language, suppose that **B** is the set of sentences of the propositional calculus (with the customary operations of negation, disjunction, and conjunction), and suppose that g is an arbitrary mapping that assigns a sentence (an element of **B**) to each propositional variable. It follows from the definition of sentences that, under these circumstances, there exists a unique endomorphism f of **B** that agrees with g on the propositional variables. This short argument shows that the propositional calculus behaves like a free B-algebra. In fact, **B** is a free B-algebra on a countably infinite set of generators, namely the set of propositional variables.

The substitution rule asserts that the Boolean congruence **F** associated with the propositional calculus is invariant under the endomorphism f. Precisely: if

$$p \equiv q\,(\mathbf{F}),$$

then

$$f(p) \equiv f(q)\,(\mathbf{F}).$$

If we view **F** as the unit element of the quotient (that is, as the set of elements provably equivalent to some element of the form $q \vee \neg q$), then we can view the substitution rule as asserting that **F** is closed under f: if $p \in \mathbf{F}$, then $f(p) \in \mathbf{F}$. In other words, if p is provable, then so is $f(p)$.

Every endomorphism of **B** is the extension of a mapping of propositional variables to sentences. Indeed, given such an endomorphism f, consider the restriction g of f to the set of propositional variables; then f is obviously the unique extension of g to an endomorphism of **B**. Therefore the substitution rule asserts that **F** (provable equivalence) is invariant under all the endomorphisms of **B**. This situation has its analogues in group theory and ring theory, and in those analogues the phrase used to describe it is "fully invariant". In that language, then, the Boolean congruence **F** of provable equivalence is fully invariant in the B-algebra **B** of all sentences.

In the customary formulation of the definition of the set of provable sentences, here identified with **F**, the substitution rule is part of the definition: **F** is defined as the least set of sentences including certain specified ones (the axioms) and invariant under modus ponens and the substitution rule. In our approach to the relation **F**, we dealt not with specified sentences but with specified sets of sentences (the axiom schemata), sets that were invariant under the substitution rule, and therefore we could omit the requirement of substitution invariance — it was automatically satisfied.

37. Boolean logics

We are now ready to give an algebraic approach to the propositional calculus. A *Boolean logic* (or a *propositional logic*) is a pair (**B**, **F**), where **B** is a B-algebra and **F** is a fully invariant Boolean congruence in **B**. Some important special cases of this concept are the propositional calculus itself (**B** is the set of all sentences and **F** the relation of provable equivalence), and the ones in which **B** is a Boolean algebra and **F** is a (fully invariant) Boolean congruence (filter) in it. It is important to distinguish between a Boolean logic, which is any instance of a certain kind of algebraic structure, and the propositional calculus, which is a quite concrete instance of a Boolean logic.

A Boolean logic $(\mathbf{B}, \mathbf{F})$ is *consistent* if the congruence **F** is not too large; precisely speaking, $(\mathbf{B}, \mathbf{F})$ is consistent if there is more than one equivalence class. That the logic of the propositional calculus is consistent in this sense, we have seen before. A consistent Boolean logic $(\mathbf{B}, \mathbf{F})$ is *complete* if the congruence **F** is not too small; precisely speaking, $(\mathbf{B}, \mathbf{F})$ is complete if every enlargement of **F** would render the logic inconsistent, or, equivalently, if **F** is *maximal* among the proper fully invariant Boolean congruences in **B**. We proceed to prove that the propositional calculus is complete.

There are at least two important notions of consistency and completeness. To distinguish between them, the ones defined just above are usually called *syntactic*, and the others (to be defined presently) are called *semantic*. The idea is that the concepts above are defined in terms of the internal structure, the grammar, the *syntax* of the logic; the others depend on the interpretation, the meaning, the *semantics* of logic. (Warning: the semantic concepts about to be defined exhibit their full richness in certain generalizations of Boolean logics and algebras; in the present situation they have the deceptive simplicity of a degenerate case.)

Define an *interpretation* of a Boolean algebra **A** as a homomorphism f from **A** into **2**. More generally, an interpretation of a Boolean logic $(\mathbf{B}, \mathbf{F})$ is a homomorphism f from $(\mathbf{B}, \mathbf{F})$ to $(\mathbf{2}, \mathbf{E})$, where **E** is the congruence relation of equality on **2** (the only proper congruence relation on **2**). In other words, f is a homomorphism from **B** to **2**, and

$$\text{if} \quad p \equiv q\,(\mathbf{F}), \quad \text{then} \quad f(p) \equiv f(q)\,(\mathbf{E}),$$

that is, $f(p) = f(q)$. This last condition can be expressed by the phrase "f sends **F** to **E**". The requirement that f sends **F** to **E** is equivalent to the requirement that if p is in the filter (associated with) **F**, then $f(p) = 1$. To check this, assume that f is a **2**-valued homomorphism on **B**. If f sends **F** to **E** and if p is in **F**, then $p \equiv q \vee \neg q\,(\mathbf{F})$ for some q, and therefore

$$f(p) = f(q \vee \neg q) = f(q) \vee \neg f(q) = 1.$$

Conversely, if f maps every element in the filter to 1, and if $p \equiv q\,(\mathbf{F})$, then the element $p \Leftrightarrow q$ is in $\mathbf{F}$, so

$$1 = f(p \Leftrightarrow q) = (f(p) \Leftrightarrow f(q)),$$

and therefore $f(p) = f(q)$.

The notion of an interpretation of $(\mathbf{B}, \mathbf{F})$ can equivalently be defined as a homomorphism from the quotient algebra $\mathbf{B}/\mathbf{F}$ onto $\mathbf{2}$. Indeed, if f is a homomorphism from $(\mathbf{B}, \mathbf{F})$ to $(\mathbf{2}, \mathbf{E})$, then define g on $\mathbf{B}/\mathbf{F}$ by setting $g([p]) = f(p)$ for each equivalence class $[p]$ of $\mathbf{F}$. The function g is well defined because f sends $\mathbf{F}$ to $\mathbf{E}$, and it is easy to check that g is a homomorphism onto $\mathbf{2}$. Conversely, suppose that a function g maps $\mathbf{B}/\mathbf{F}$ homomorphically onto $\mathbf{2}$. Define a function f on $\mathbf{B}$ by setting $f(p) = g([p])$. It is a straightforward matter to check that f is a homomorphism from $(\mathbf{B}, \mathbf{F})$ to $(\mathbf{2}, \mathbf{E})$.

An automatic verification shows that this general definition of interpretation contains the earlier one, for the propositional calculus, as a special case. An element p of $\mathbf{A}$ (or of $\mathbf{B}$) is *valid* if it is *true* (mapped to 1) in every interpretation; if p is *false* (mapped to 0) in every interpretation, then p is called *contravalid*. Valid and contravalid elements are usually called *tautologies* and *contradictions* respectively. An element p of $\mathbf{A}$ (or $\mathbf{B}$) is *satisfiable* if there is an interpretation for which it is not false (that is, if p is not contravalid).

A Boolean logic is *semantically consistent* if it has an interpretation. The definition of semantic completeness requires one more sentence of motivation. It is part of the definition of interpretation (for a Boolean logic $(\mathbf{B}, \mathbf{F})$) that the "unit element" (that is, any element p of $\mathbf{B}$ that is congruent modulo $\mathbf{F}$ to one of the form $q \vee \neg q$) is always valid. Semantic completeness requires the converse: $(\mathbf{B}, \mathbf{F})$ is *semantically complete* if "the unit" is the only valid element. More explicitly, semantic completeness requires that if p (in $\mathbf{B}$) is valid, then p must be congruent modulo $\mathbf{F}$ to an element of the form $q \vee \neg q$. (Of course, all elements of the form $q \vee \neg q$ are congruent to one another modulo $\mathbf{F}$ because $\mathbf{F}$ is a Boolean congruence. Thus, it is not necessary to specify a particular q.)

The syntactic consistency of the propositional calculus was proved earlier by proving its semantic consistency. In fact, semantic consistencyimplies syntactic consistency for any Boolean logic $(\mathbf{B}, \mathbf{F})$: if f maps $\mathbf{B}$ onto $\mathbf{2}$, so that the elements $q \vee \neg q$ get mapped to 1 and their negation to 0, then $\mathbf{B} \neq \mathbf{F}$. We shall now prove the syntactic completeness of the propositional calculus by pointing out its semantic completeness.

Semantic completeness theorem. *In a Boolean logic* $(\mathbf{B}, \mathbf{F})$, *if* p *is a valid sentence, then*

$$p \equiv q \vee \neg q\,(\mathbf{F}).$$

Proof. The semantic completeness theorem is really just another version of the existence theorem. To show this, we write a series of equivalent formulations. The contrapositive of the theorem is:

$$\text{if} \quad p \not\equiv q \vee \neg q\,(\mathbf{F}), \quad \text{then } p \text{ is not valid.}$$

Because the theorem is making an assertion about all elements p of $\mathbf{B}$, and because the mapping that takes each element of $\mathbf{B}/\mathbf{F}$ to its complement is a bijection of $\mathbf{B}/\mathbf{F}$, we may replace "p" in the contrapositive by its complement:

$$\text{if} \quad \neg p \not\equiv q \vee \neg q\,(\mathbf{F}), \quad \text{then } \neg p \text{ is not valid.}$$

In terms of the quotient algebra $\mathbf{B}/\mathbf{F}$, this says that if $[\neg p] \neq 1$, then there is a $\mathbf{2}$-valued homomorphism f on $\mathbf{B}/\mathbf{F}$ such that $f([\neg p]) \neq 1$. Equivalently, if $[p] \neq 0$, then there is a $\mathbf{2}$-valued homomorphism f on $\mathbf{B}/\mathbf{F}$ such that $f([p]) = 1$. This last statement is a direct consequence of the existence theorem. In fact, it is equivalent to the existence theorem, since every Boolean algebra $\mathbf{B}$ can be written as (is isomorphic to) the quotient $\mathbf{B}/\mathbf{F}$, where $\mathbf{F}$ is the (invariant) identity congruence on $\mathbf{B}$. The proof is complete.

How does it follow that the *propositional calculus* is syntactically complete? Let $\mathbf{B}$ be the propositional calculus and $\mathbf{F}$ any proper fully invariant Boolean congruence on $\mathbf{B}$ (for example, the relation of provable equivalence). To show that $\mathbf{F}$ must be a maximal fully invariant congruence, consider an element p of $\mathbf{B}$ whose equivalence class $[p]$ is *not* the unit element of the quotient algebra $\mathbf{B}/\mathbf{F}$. It is to be proved that if $\tilde{\mathbf{F}}$ is a fully invariant Boolean congruence that includes $\mathbf{F}$ and "contains" p, that is, if

$$p \equiv q \vee \neg q\,(\tilde{\mathbf{F}}),$$

then $\tilde{\mathbf{F}} = \mathbf{B}$. Since $[p]$ is not the unit element, it follows from the contrapositive of the semantic completeness theorem that there is a $\mathbf{2}$-valued homomorphism f on $\mathbf{B}/\mathbf{F}$ such that $f([p]) = 0$. Let q_0 and q_1 be elements of $\mathbf{B}$ such that $[q_0] = 0$ and $[q_1] = 1$ (in $\mathbf{B}/\mathbf{F}$). For each propositional variable v, write $g(v) = q_0$ or q_1 according as $f([v]) = 0$ or 1. Since $\mathbf{B}$ is freely generated by the propositional variables (recall that it is the propositional calculus), there exists a unique endomorphism $\tilde{g}$ of $\mathbf{B}$ that agrees with g on the set of propositional variables. Since $\tilde{\mathbf{F}}$ is fully invariant and "contains" p, it "contains" $\tilde{g}(p)$.

Next we show that $\tilde{g}(p)$ is contravalid in the propositional calculus, that is, in $(\mathbf{B}, \mathbf{F})$. To this end, let h be an arbitrary $\mathbf{2}$-valued homomorphism on $\mathbf{B}/\mathbf{F}$ and k the canonical homomorphism from $\mathbf{B}$ onto $\mathbf{B}/\mathbf{F}$. Fix any propositional variable v. If $g(v) = q_0$, then

$$\begin{aligned}(h \circ k \circ \tilde{g})(v) &= h([\tilde{g}(v)]) = h([g(v)]) = h([q_0]) \\ &= h(0) = 0 = f([v]) = (f \circ k)(v).\end{aligned}$$

Similarly, if $g(v) = q_1$, then

$$\begin{aligned}(h \circ k \circ \tilde{g})(v) &= h([\tilde{g}(v)]) = h([g(v)]) = h([q_1]) \\ &= h(1) = 1 = f([v]) = (f \circ k)(v).\end{aligned}$$

Since the homomorphisms $h \circ k \circ \tilde{g}$ and $f \circ k$ agree on the propositional variables, and since the propositional variables generate $\mathbf{B}$, it follows that the homomorphisms agree on all sentences in $\mathbf{B}$. In particular,

$$h([\tilde{g}(p)]) = (h \circ k \circ \tilde{g})(p) = (f \circ k)(p) = f([p]) = 0.$$

Since h was an arbitrary interpretation, $\tilde{g}(p)$ must be contravalid, and therefore $\neg \tilde{g}(p)$ is valid.

By the semantic completeness theorem, we conclude that $[\neg \tilde{g}(p)] = 1$ (in $\mathbf{B}/\mathbf{F}$). In other words, $\neg \tilde{g}(p)$ is "in" $\mathbf{F}$. Because $\tilde{\mathbf{F}}$ includes $\mathbf{F}$, it follows that $\neg \tilde{g}(p)$ is "in" $\tilde{\mathbf{F}}$ as well. But $\tilde{\mathbf{F}}$ "contains" $\tilde{g}(p)$. Therefore $\tilde{\mathbf{F}} = \mathbf{B}$.

38. Algebra of the propositional calculus

Everything of logical interest that can be said about the sentences of the propositional calculus is invariant under (compatible with) provable equivalence. In what follows we shall, therefore, concentrate on the algebra of the propositional calculus, rather than its combinatorics; practically speaking, this means that we shall study certain elementary aspects of the theory of Boolean algebras. The selection of the particular aspects to be studied is motivated by logical considerations.

It is sometimes convenient to be able to speak of "an equivalence class of the propositional calculus with respect to provable equivalence" without being quite so wordy; we propose (as mentioned before) to do so by use of the word "proposition." A proposition, therefore, is an element of a free Boolean algebra with countably many generators; at times, it will be convenient to use the word for an arbitrary element of a quite arbitrary Boolean algebra.

An important role is often given to the topic of "normal forms" for propositions. The question that they answer can, roughly speaking, be asked this way: what do propositions look like? More precisely: if **A** is the free Boolean algebra on a set **E** of generators, how can the elements of **A** be expressed in terms of those of **E**? One answer is this: every element of **A** is a finite join (supremum, disjunction) of finite meets (infima, conjunctions); each term in those meets is either an element of **E** or the complement (negation) of one. More symbolically: every element of **A** is the supremum of finitely many elements of the form

$$\varepsilon_1 p_1 \wedge \varepsilon_2 p_2 \wedge \cdots \wedge \varepsilon_n p_n,$$

where $p_i \in \mathbf{E}$ $(i = 1, \ldots, n)$, and where each ε_i is either the empty symbol (in which case $\varepsilon_i p_i$ is p_i) or the symbol for complementation (in which case $\varepsilon_i p_i$ is $\neg\, p_i$).

The proof of this assertion is straightforward (and it does not even depend on the assumption that the elements of **E** are *free* generators of **A**): just observe that the set **S** of elements so described constitutes a Boolean subalgebra of **A** that contains **E** and therefore coincides with **A**. Indeed, it is clear that **S** is closed under $\vee$. The set **S** is also closed closed under $\wedge$. The proof of this assertion consists of just "multiplying out", using the distributive law. Here is an example:

$$\begin{aligned} &((p \wedge \neg q) \vee (\neg r \wedge s)) \wedge (r \vee (s \wedge \neg t)) \\ &\quad = ((p \wedge \neg q) \wedge (r \vee (s \wedge \neg t))) \vee ((\neg r \wedge s) \wedge (r \vee (s \wedge \neg t))) \\ &\quad = (p \wedge \neg q \wedge r) \vee (p \wedge \neg q \wedge s \wedge \neg t) \vee (\neg r \wedge s \wedge r) \vee (\neg r \wedge s \wedge s \wedge \neg t). \end{aligned}$$

By the De Morgan laws, the complement of a meet term

$$\varepsilon_1 p_1 \wedge \varepsilon_2 p_2 \wedge \cdots \wedge \varepsilon_n p_n$$

is the join

$$\delta_1 p_1 \vee \delta_2 p_2 \vee \cdots \vee \delta_n p_n$$

of monomials $\delta_i p_i$, where δ_i is empty if ε_i is $\neg$, and δ_i is $\neg$ if ε_i is empty. Since each monomial is, by definition, a meet term, the join of such monomials must be in **S**. Thus, the complement of every meet term in **S** is again in **S**. The complement of a join of such meet terms is the meet of their complements, by the De Morgan laws. Since **S** is closed under meets and contains the complements of its meet terms, it follows that **S** must be closed under complementation. The proof is complete.

Note that it is not asserted that this representation (or normal form) is unique, and, indeed, it is not: witness

$$p \vee \neg p = q \vee \neg q.$$

The normal form we have just described is called *disjunctive normal form*; its dual (in an obvious sense) is called *conjunctive normal form*.

39. Algebra of proof and consequence

We turn next to the algebraic formulation of the concepts of proof and consequence. A proof in the propositional calculus was a finite sequence of sentences such that each sentence in the sequence is either (i) an axiom, or (ii) the result of applying modus ponens to some pair of preceding sentences in the sequence. In the algebraic version of logic this concept no longer makes sense, but a slight generalization of it does. It does not make sense as is because there are no more "axioms" — the sentences that were called that were swallowed up by the set of all theorems and collectively identified with the unit element of the algebra. The generalization that makes sense is that of "proof from hypotheses".

If **S** is a set of sentences of the propositional calculus, a proof from the hypotheses **S** is a finite sequence of sentences such that each element in the sequence is either

(i) an axiom or else in **S**,

or

(ii) the result of applying modus ponens to some pair of preceding sentences in the sequence.

The algebraic version is the same except that it replaces "sentence" by "proposition" and eliminates "axiom". Thus, if **A** is a Boolean algebra and **S** is a subset of **A**, a proof from the hypotheses **S** is a finite sequence of propositions (elements of **A**) such that each proposition in the sequence is either

(i) equal to 1 or else in **S**,

or

(ii) the result of applying modus ponens to some pair of preceding propositions in the sequence.

The last term of a proof is said to be a *consequence* of the hypotheses **S**. (Modus ponens, incidentally, means the same algebraically as combinatorially; it is the ternary relation that holds for p, q, and r just in case $q = (p \Rightarrow r)$. As for $p \Rightarrow r$, that makes good algebraic sense: it is, of course, the proposition $\neg p \vee r$.)

What can be said about the set $\hat{\mathbf{S}}$ of all consequences of a set **S**? If **S** is empty, then $\hat{\mathbf{S}}$ consists of 1 alone; this is clear. Since more generally, 1 is always in $\hat{\mathbf{S}}$, and since

$$\neg p \vee \neg q \vee (p \wedge q) = 1$$

for all p and q, it follows that if p is in $\hat{\mathbf{S}}$, then so is $q \Rightarrow (p \wedge q)$. (Note: $(\neg p \vee \neg q \vee (p \wedge q)) = (p \Rightarrow (q \Rightarrow (p \wedge q)))$.) From this, in turn, it follows that if both p and q are in $\hat{\mathbf{S}}$, then so is $p \wedge q$. If, on the other hand, p is in $\hat{\mathbf{S}}$ and q is an element of **A** such that $p \leq q$, then

$$p \Rightarrow q = \neg p \vee q \geq \neg p \vee p = 1,$$

so that $p \Rightarrow q$ is in $\hat{\mathbf{S}}$, and therefore q is in $\hat{\mathbf{S}}$. Conclusion: $\hat{\mathbf{S}}$ is a filter.

What was just proved is that the set $\hat{\mathbf{S}}$ of all consequences of **S** is a filter including **S**; it is, in fact, the filter generated by **S**. To prove this we must prove that if **F** is a filter such that $\mathbf{S} \subset \mathbf{F}$, then $\hat{\mathbf{S}} \subset \mathbf{F}$. For this purpose, however, it is sufficient to prove that **F** is closed under modus ponens, that is, that if p and $p \Rightarrow q$ are in **F**, then so is q. This is trivial:

$$p \wedge (p \Rightarrow q) = p \wedge (\neg p \vee q) = p \wedge q \leq q.$$

Conclusion: the set of all consequences of every set **S** is the filter generated by **S**.

From this conclusion, and from the characterization of the filter generated by a set, we immediately draw the following corollary: q is a consequence of **S** if and only if there is a finite sequence $p_1, \ldots, p_n$ of elements of **S** such that

$$p_1 \wedge \cdots \wedge p_n \leq q,$$

that is, such that the element

$$(p_1 \wedge \cdots \wedge p_n) \Rightarrow q$$

is the unit of **A**.

There is a further generalization of this generalization of proof that is often useful, namely, the generalization to Boolean logics. Suppose that **A** is a B-algebra and **F** is a Boolean congruence in **A**; a proof from the hypotheses **S** in (**A**, **F**) is a finite sequence of elements of **A** such that each proposition in the sequence is either (i) "in" **F** or in **S**, or (ii) the result of applying modus ponens to some pair of preceding propositions in the sequence. It follows just as above that in this context the set of all consequences of a set **S** is the filter generated by **S** and (the filter associated with) **F**. In particular, the set of all consequences of the empty set in (**A**, **F**) is just **F** itself.

Here is one standard (logical) notation associated with these ideas. If (**A**, **F**) is a Boolean logic, and if **S** and **T** are subsets of **A**, write

$$\mathbf{S} \vdash \mathbf{T}$$

to mean that **T** is included in the set of all consequences of **S**, that is, in the filter generated by **S** and **F**. If **S** is empty, then do not write it (that is, $\vdash \mathbf{T}$ means that $\mathbf{T} \subset \mathbf{F}$); if **T** is empty, nothing is being said. If either **S** or **T** is a singleton $\{p\}$, it is customary to replace the symbol for **S** or for **T**, as the case may be, by "p" (not "$\{p\}$"); for example, $p \vdash \neg p \vee p$.

The apparently more general case of Boolean logics (**A**, **F**) is not really more general than the special case of Boolean algebras ($\mathbf{F} = \{1\}$); all that is necessary to recapture the general theory from the special case is to pass from **A** to **A**/**F**. Consider, for instance, the following assertion.

Deduction theorem. $\mathbf{S} \vDash q$ *if and only if* **S** *has a finite subset* $\{p_1, \ldots, p_n\}$ *such that* $\vdash (p_1 \wedge \cdots \wedge p_n) \Rightarrow q$.

In view of the remark about **A/F**, it is sufficient to prove the result in case $\mathbf{F} = \{1\}$. The result in that case takes the form

$\mathbf{S} \vDash q$ if and only if $p_1 \wedge \cdots \wedge p_n \leq q$ for some finite subset $\{p_1, \ldots, p_n\}$ of **S**.

The last assertion follows from the fact that the set of consequences of **S** coincides with the filter generated by **S**.

One important special case of the deduction theorem is that $p \vDash q$ if and only if $\vdash p \Rightarrow q$ (that is, if and only if $p \leq q$).

A standard observation about the propositional calculus, called the (*weak*) *soundness theorem*, asserts (in the context of a Boolean logic $(\mathbf{A}, \mathbf{F})$) that every provable proposition (that is, every consequence of the empty set) is valid. Since the provable propositions are just the elements of $\mathbf{F}$, this theorem is an immediate consequence of the definition of an interpretation of the logic $(\mathbf{A}, \mathbf{F})$. To formulate a generalization of the soundness theorem, we need to introduce the relation of semantic consequence. A proposition q is a *semantic consequence* in $(\mathbf{A}, \mathbf{F})$ of a set $\mathbf{S}$ of propositions if every interpretation of $(\mathbf{A}, \mathbf{F})$ that assigns the value 1 to each of the elements of $\mathbf{S}$ must also assign the value 1 to q. The notation that is often employed to express relationship of semantic consequence between $\mathbf{S}$ and q is $\mathbf{S} \vDash q$. As usual, when $\mathbf{S}$ is empty we don't write it, that is, we write $\vDash q$. The notation $\vDash q$ expresses that q is valid in $(\mathbf{A}, \mathbf{F})$.

The soundness and semantic completeness theorems for $(\mathbf{A}, \mathbf{F})$ can be succinctly stated using the notation we have introduced. Soundness says: if $\vdash q$, then $\vDash q$. Semantic completeness is just the converse: if $\vDash q$, then $\vdash q$. Strong soundness and strong completeness are the generalizations of these statements where the (invisible) empty set on the left is replaced by an arbitrary set $\mathbf{S}$ of propositions. We shall formulate them for an arbitrary Boolean logic $(\mathbf{A}, \mathbf{F})$. Interestingly, the strong versions of soundness and completeness are actually simple consequences of the weak versions.

Strong soundness theorem. *For any proposition q and any set $\mathbf{S}$ of propositions in $(\mathbf{A}, \mathbf{F})$, if $\mathbf{S} \vdash q$, then $\mathbf{S} \vDash q$.*

Proof. Suppose that $\mathbf{S} \vdash q$. The deduction theorem says that there is a finite subset $\{p_1, \ldots, p_n\}$ of $\mathbf{S}$ such that

$$\vdash (p_1 \wedge \cdots \wedge p_n) \Rightarrow q.$$

By the (weak) soundness theorem, this gives

$$\vDash (p_1 \wedge \cdots \wedge p_n) \Rightarrow q;$$

in other words, the proposition $(p_1 \wedge \cdots \wedge p_n) \Rightarrow q$ is valid. In particular, any interpretation of $(\mathbf{A}, \mathbf{F})$ that assigns the value 1 to each of $p_1, \ldots, p_n$ will assign the value 1 to q. Thus, $\mathbf{S} \vDash q$.

Strong completeness theorem. *For any proposition q and any set $\mathbf{S}$ of propositions in $(\mathbf{A}, \mathbf{F})$, if $\mathbf{S} \vDash q$, then $\mathbf{S} \vdash q$.*

Proof. Suppose that $\mathbf{S} \models q$. Let $\hat{\mathbf{S}}$ be the set of (provable) consequences of $\mathbf{S}$. Then $\hat{\mathbf{S}}$ is a filter that includes $\mathbf{F}$. We must show that q is in $\hat{\mathbf{S}}$. The assumption $\mathbf{S} \models q$ implies that in the Boolean logic $(\mathbf{A}, \hat{\mathbf{S}})$ the proposition q is valid. Therefore q is provable (in symbols, $\vdash q$) in $(\mathbf{A}, \hat{\mathbf{S}})$ by the (weak) completeness theorem for $(\mathbf{A}, \hat{\mathbf{S}})$. In other words, $q \in \hat{\mathbf{S}}$, as was to be shown.

There is a surprising and important semantic analogue of the deduction theorem.

Compactness theorem. *For any proposition q and any set $\mathbf{S}$ of propositions in $(\mathbf{A}, \mathbf{F})$, we have $\mathbf{S} \models q$ if and only if there is a finite subset $\{p_1, \ldots, p_n\}$ of $\mathbf{S}$ such that $\{p_1, \ldots, p_n\} \models q$.*

To prove the theorem, just apply the strong soundness theorem, the strong completeness theorem, and the deduction theorem.

There is much that could still be said about the propositional calculus and its generalizations, but we shall not do so. We propose to stop short the study of the propositional calculus right here, and to go on to the algebraic study of a deeper, more important, and, incidentally, older branch of logic, the theory of syllogisms. Before taking up that study, we pause briefly to discuss some notions that generalize the concept of a Boolean algebra and the concepts of a finite join and meet of elements.

Lattices and infinite operations

40. Lattices

There are subjects other than the ones discussed above that are sometimes classified under the phrase "propositional calculus". We mean not only that there are further theorems of interest that could have been included; we mean that there are further theories that are sufficiently similar to bear the same name (sometimes).

One way to unify the various propositional theories is to introduce the concept of a *lattice*, a generalization of Boolean algebra. We have seen that in a Boolean algebra there is a natural order relation $\leq$. The order is such that, corresponding to any two elements p and q, there is a (necessarily unique) smallest element r with the property that $p \leq r$ and $q \leq r$, and, similarly, corresponding to any two elements p and q there is a (necessarily unique) largest element s with the property that $s \leq p$ and $s \leq q$. These elements (known as the *supremum* or *join*, and the *infimum* or *meet*, of p and q respectively) are denoted by $p \vee q$ and $p \wedge q$ respectively. The concept of lattice generalizes this aspect of the theory of Boolean algebras: it is a partially ordered set in which each pair of elements has a supremum and an infimum. Just as their Boolean counterparts, the join and meet operations of a lattice are associative, commutative, and idempotent.

A lattice is *complemented* if (i) there exist in it elements 0 and 1 such that $0 \leq p \leq 1$ for all p, and (ii) to each element p in it there corresponds at least one element q such that $p \wedge q = 0$ and $p \vee q = 1$. A lattice is *distributive* if

$$p \wedge (q \vee r) = (p \wedge q) \vee (p \wedge r)$$

for all p, q, and r. Alternatively, a lattice could be called distributive if

$$p \vee (q \wedge r) = (p \vee q) \wedge (p \vee r);$$

fortunately, it turns out that these two identities are equivalent. Another fortunate circumstance is that in a distributive lattice complementation is unique; that is, if

$$p \wedge q = p \wedge r = 0 \quad \text{and} \quad p \vee q = p \vee r = 1,$$

then

$$q = r.$$

Boolean algebras can be characterized as complemented distributive lattices.

The other kinds of lattices that enter logic are associated with such phrases as "intuitionistic logic", "modal logic", "many-valued logic". The lattices of intuitionistic logic are obtained from the theory of Boolean algebras by dropping conditions; the lattices of modal logic are obtained by adding conditions. (An oblique view of modal logics will appear in our subsequent treatment of syllogisms via monadic algebras.) The lattices of many-valued logics are the weirdest of all, but they are mildly amusing, and it is so easy to describe them that we take a minute to do so.

A field **B** of subsets of a set X is a collection of subsets of X closed under the set-theoretic operations of union, intersection, and complementation. If, instead of the collection **B** of subsets of X, we consider the corresponding collection of their characteristic functions, we obtain a notationally different description of what is conceptually obviously the same thing. In terms of characteristic functions, the operations that a field is closed under can be described as follows: the join of two characteristic functions p and q is the function $p \vee q$ defined by

$$(p \vee q)(x) = \max\,(p(x), q(x))\,;$$

their meet is the function $p \wedge q$ defined by

$$(p \wedge q)(x) = \min\,(p(x), q(x))\,;$$

and the complement of p is the function $\neg\, p$ defined by

$$(\neg\, p)(x) = p(x) + 1 \pmod 2.$$

The set of all functions from X into $\mathbf{2}$ (that is, the set of characteristic functions of subsets of X) is usually denoted by $\mathbf{2}^X$, and this same notation is used to denote the corresponding Boolean algebra of functions. In this notation Stone's representation theorem says that every Boolean algebra is isomorphic to a subalgebra of $\mathbf{2}^X$ for some set X.

Suppose now that 2 is replaced by some other integer n (> 1). That is, given X, consider functions on X whose values are the numbers $0, 1, \ldots, n-1$, and define for them the operations

$$p \vee q \,(\max),$$

$$p \wedge q \,(\min),$$

and

$$\neg\, p \,(\text{successor modulo } n),$$

and consider, in particular, a class of such functions closed under the operations so described. Much of the study of many-valued logics concentrates on this structure and its various possible generalizations and representations.

41. Non-distributive lattices

We have seen that some lattices are distributive; we have not yet seen that some are not. Let us digress to a small fragment of lattice theory, just for fun. We do not propose ever in these notes to tie it up with logic.

Some lattices have only a finite number of elements. For instance, the Boolean algebra $\mathbf{2}$ is a lattice with just two elements, 0 and 1. It is customary and convenient to indicate the order relation in a finite lattice by a diagram (sometimes called the *Hasse diagram*) in which each element is represented by a dot, each element *greater* than a given is drawn *above* the given one, and two elements p and q, with $p \leq q$ and $p \neq q$, are joined by a line segment just in case there is no third element between them. Thus the diagram of $\mathbf{2}$ is

The diagram of the lattice (Boolean algebra) of all subsets of 2 is

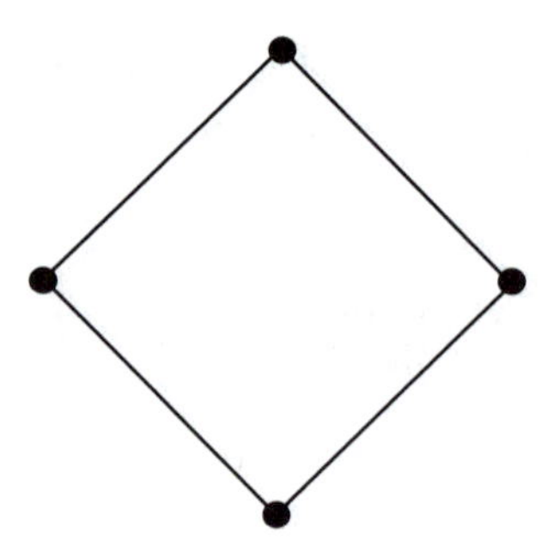

The diagram of the lattice of all subsets of a three-element set is

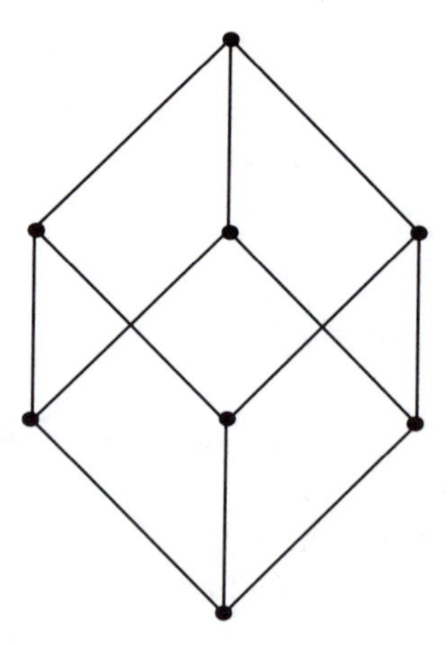

(Note that the intersction of two segments need not be a dot.)

When is a randomly drawn diagram the Hasse diagram of a (finite) lattice? Every diagram indicates a (partial) order relation among its dots: $p \leq q$ if and only if either $p = q$ or there is an upward path of segments from p to q. Thus

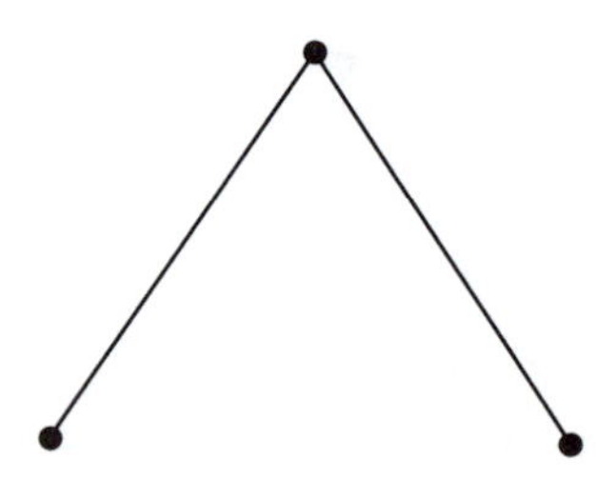

is the diagram of something, but the something is surely not a lattice: the two lower points have no infimum. A necessary condition that a diagram be the Hasse diagram of a lattice is that each pair of dots be joinable to some dot that is above both and to some dot that is below both. (In lattice language this requires the existence of upper and lower bounds for each pair of elements.) This condition is not sufficient, however. Example: the diagram

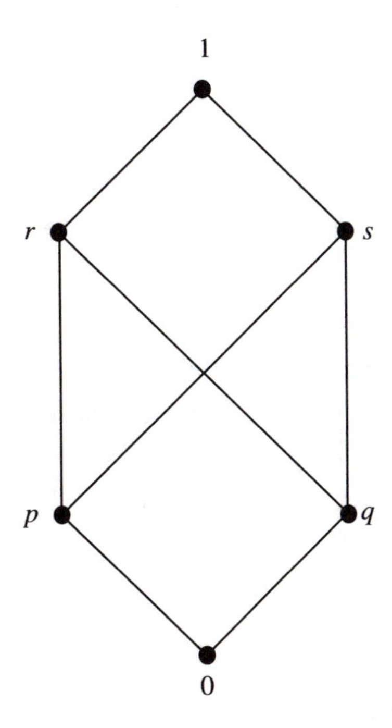

is not the diagram of a lattice. Reason: p and q have no *least* upper bound. There does not seem to be any satisfactory answer to the question of which diagrams are the diagrams of lattices; the only thing to do is to check the existence of all suprema and infima.

Note: in every lattice, finite or not, if there exists an element p, necessarily unique, such that $p \le q$ for all q, then p is called the *zero* of the lattice, and is denoted by 0; similarly if $q \le p$ for all q, then p is called the *unit* of the lattice, and is denoted by 1. Infinite lattices may fail to possess a zero and a unit; finite ones clearly always do have them.

It is easy to check that the lattices whose diagrams were given above (obviously we cannot here mean the non-lattices whose diagrams were also given above) are distributive. Consider, on the other hand, the diagram that follows.

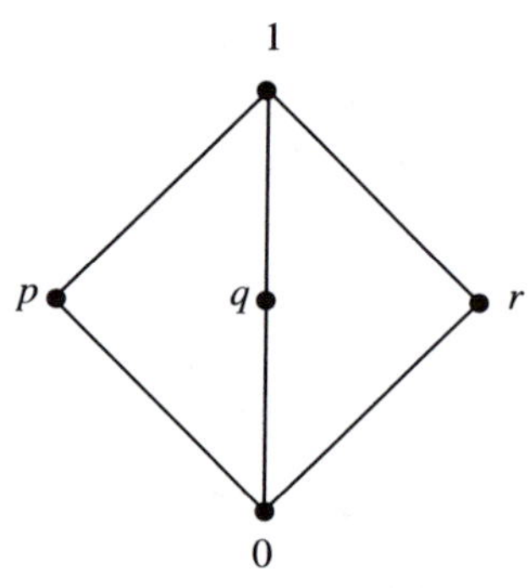

Assertion: it is the diagram of a non-distributive lattice. That it is a lattice is a routine verification. To prove that it is not distributive, verify that

$$p \wedge (q \vee r) = p \wedge 1 = p \quad \text{and} \quad (p \wedge q) \vee (p \wedge r) = 0 \vee 0 = 0.$$

A lattice is called *modular* if a fragment of the distributive law,

$$p \wedge (q \vee r) = (p \wedge q) \vee (p \wedge r),$$

holds in it; the appropriate fragment consists of all those instances of this distributive law for which $r \leq p$. Under this assumption, the element $p \wedge r$ coincides with r. Therefore, the equation can be rewritten in the simpler form

$$p \wedge (q \vee r) = (p \wedge q) \vee r,$$

where $r \leq p$. A discussion of the reason for the importance of the modular identity would be too great a digression for us now. We content ourselves with some trivial observations. First: modularity is genuinely weaker than distributivity. Example: the non-distributive lattice given above is modular. Second: not every lattice is modular. Example:

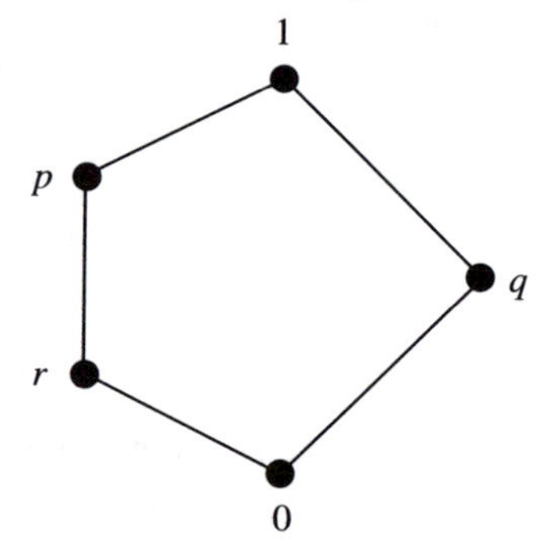

Here

$$p \wedge (q \vee r) = p \wedge 1 = p \quad \text{and} \quad (p \wedge q) \vee r = 0 \vee r = r.$$

Recall that the ordinary (arithmetic) distributive law,

$$a(b + c) = ab + ac,$$

is verbally described by saying that multiplication is distributive over addition. If, more generally, X is any set with a unary operation A (which makes correspond to each element b an element $A(b)$) and a binary operation $\to$ (which makes correspond to each pair of elements a and b an element $a \to b$), then we might say that A is distributive over $\to$ if

$$A(a \to b) = A(a) \to A(b)$$

for all a and b. (In this language, then, the verbal description of the arithmetic distributive law becomes, in pedantically detailed form, the following statement: for each number a, the unary operation of left multiplication by a is distributive over addition.) The correspondingly generalized version of the modular law is the requirement that

$$A(a \to b) = A(a) \to b,$$

not quite for all a and b, but only for those for which b is invariant under A, that is, $A(b) = b$; in that case, of course, the equation can also be made to take the form of a distributive law. (Note that in the form displayed above it is a kind of associative law.)

The preceding discussion may serve as a mnemonic device for the lattice-theoretic modular law. That law says that, for each element p, the unary operation of p-infimum formation, that is, the operation

$$q \mapsto p \wedge q,$$

is modular over supremum. Indeed, the invariance ($b = A(b)$) becomes in this case $r = p \wedge r$, or $r \leq p$, and, consequently, modularity becomes just what it was originally defined to be.

42. Infinite operations

In a lattice every two-element set has a supremum (by definition), or, to put it more exactly, the set of all upper bounds of a two-element set is not empty and has a (necessarily unique) least element. A routine induction argument implies that once this is assumed for two-element sets, the same result holds for all finite sets. For infinite sets, however, the situation is different, even in Boolean algebras. The trouble is not with the existence of upper bounds (since Boolean algebras have unit elements, every set in a Boolean algebra is bounded), but with the existence of a least element in the set of all upper bounds. Consider, for an example, the set X of all positive integers, and the field (Boolean algebra) of all those subsets of X that either are finite or have finite complements. The set of all singletons consisting of odd numbers, that is, the set $\{\{1\},\{3\},\{5\},\ldots\}$, has many upper bounds. The complement of each finite set of even numbers is such an upper bound, and, in fact, every upper bound has that form. Since there is no largest finite set of even numbers, there is no smallest set whose complement is a finite set of even numbers. In other words, the set $\{\{1\},\{3\},\{5\},\ldots\}$ has no least upper bound. This example shows the possible non-existence of suprema for infinite sets in Boolean algebras.

Although infinite suprema do not always exist, they do have a theory. To formulate some simple laws involving them, it is helpful to introduce a bit of notation. If $\{p_i : i \in I\}$ is a family of elements in a Boolean algebra (or, for that matter, in a lattice), we shall often supress the index set I and write simply $\{p_i\}$. The supremum and infimum of this set, provided they exist, will be denoted by

$$\bigvee_i p_i \qquad \text{and} \qquad \bigwedge_i p_i$$

respectively.

A trivial (almost vacuous) part of that theory is the infinite commutative law, which says that infinite suprema are independent of the order of the elements used to form them. More precisely: if $\{p_i\}$ is a family of elements (in a Boolean algebra) that has a supremum, and if π is a permutation of the indices, then $\{p_{\pi(i)}\}$ has a supremum also and

$$\bigvee_i p_i = \bigvee_i p_{\pi(i)}.$$

The proof consists of the observation that the *sets* $\{p_i\}$ and $\{p_{\pi(i)}\}$ are the same and hence have the same sets of upper bounds.

The following assertion is the infinite version of the De Morgan laws. If $\{p_i\}$ is a family of elements in a Boolean algebra, then

$$\neg\left(\bigvee_i p_i\right) = \bigwedge_i \neg p_i \qquad \text{and} \qquad \neg\left(\bigwedge_i p_i\right) = \bigvee_i \neg p_i.$$

The equations are to be interpreted in the sense that if either term in either equation exists, then so does the other term of that equation, and the two terms are equal.

For the proof, suppose that $p = \bigvee_i p_i$. Since $p_i \leq p$ for every i, it follows that $\neg p \leq \neg p_i$ for every i. In other words, $\neg p$ is a lower bound for $\{\neg p_i\}$. It is to be proved that $\neg p$ is the greatest such lower bound, that is:

$$\text{if} \quad q \leq \neg p_i \quad \text{for every } i\text{, then} \quad q \leq \neg p.$$

The assumption implies that $p_i \leq \neg q$ for every i, and hence from the definition of supremum,

$$p \leq \neg q.$$

Therefore

$$q \leq \neg p.$$

A dual argument justifies the passage from the left side of the second equation to the right. To justify the reverse passages, apply the results already proved to the family of complements. For instance, suppose that $p = \bigwedge_i \neg p_i$. Then

$$\neg p = \neg \left(\bigwedge_i \neg p_i \right) = \bigvee_i \neg \neg p_i = \bigvee_i p_i,$$

and therefore

$$p = \neg \left(\bigvee_i p_i \right).$$

This completes the proof.

There is also an infinite associative law; it can be stated as follows. If $\{I_j\}$ is a family of sets with union I, and if $\{p_i : i \in I\}$ is a family of elements in a Boolean algebra, then

$$\bigvee_j \left(\bigvee_{i \in I_j} p_i \right) = \bigvee_{i \in I} p_i.$$

The equation is to be interpreted in the sense that if each of the suprema on the left exists, then so does the supremum on the right, and the two sides of the equation are equal.

For the proof, write

$$q_j = \bigvee_{i \in I_j} p_i$$

and

$$q = \bigvee_j q_j.$$

We are to prove that q is an upper bound of the family $\{p_i : i \in I\}$, and that, in fact, it is the least upper bound. Since each i in I belongs to some I_j, it follows that for each i there is a j with $p_i \leq q_j$; since, moreover, $q_j \leq q$, it follows that q is indeed an upper bound. Suppose now that $p_i \leq r$ for every i. Since, in particular, $p_i \leq r$ for every i in I_j, it follows from the definition of supremum that $q_j \leq r$. Since this is true for every j, we conclude, similarly, that $q \leq r$, and this completes the proof.

The infinite associative law implies the finite associative law. To see this, consider three elements p, q, and r. If I_0 is $\{p, q\}$ and I_1 is $\{r\}$, then the infinite associative law says that the element $(p \vee q) \vee r$ is the supremum of $\{p, q, r\}$. Similarly, if I_0 is $\{p\}$ and I_1 is $\{q, r\}$, then the element $p \vee (q \vee r)$ is the supremum of $\{p, q, r\}$. Therefore

$$(p \vee q) \vee r = p \vee (q \vee r).$$

The finite commutative law and idempotence law can be similarly obtained from the infinite associative law. The formulation of the infinite associative law can also be interpreted so as to include the infinite commutative law and the infinite idempotence law as a special case.

The preceding comments on infinite commutativity and associativity were made for suprema; it should go without saying that the corresponding (dual) comments for infima are just as true. The most interesting infinite law is the one in which suprema and infima occur simultaneously; that is the distributive law. It, too, comes in a dual pair; we shall take advantage of the principle of duality and restrict our attention to only one member of the pair.

If p is an element and $\{q_i\}$ is a family of elements in a Boolean algebra, then

$$p \wedge \bigvee_i q_i = \bigvee_i (p \wedge q_i).$$

The equation is to be interpreted in the sense that if the supremum on the left exists, then so does the one on the right, and the two sides of the equation are equal.

To prove the assertion, write $q = \bigvee_i q_i$; clearly

$$p \wedge q_i \leq p \wedge q$$

for every i. It is to be proved that if $p \wedge q_i \leq r$ for every i, then $p \wedge q \leq r$. For the proof, observe that

$$q_i = (p \wedge q_i) \vee (\neg p \wedge q_i) \leq r \vee \neg p,$$

and hence, by the definition of supremum, $q \leq r \vee \neg p$. Form the meet of both sides of this inequality with p to get $p \wedge q \leq p \wedge r$; the desired conclusion now follows from the trivial fact that $p \wedge r \leq r$.

Monadic predicate calculus

43. Propositional functions

From the point of view of the applications of logic, the propositional calculus is a pretty poor thing. It does not describe how we reason about things — it tells us merely how we reason about completed statements about things. It tells us that if we know that "all men are mortal", and if we know that "all men are mortal implies that all angels are mortal", then we can infer that "all angels are mortal". It does not tell us that if we know that "some angels are men", and if we know that "all men are mortal", then we can infer that "some angels are mortal".

The difference is big and the difference is important. The propositional calculus tosses sentences (and propositions) around like sackfuls of truth, without ever being able to look inside the sacks. It tells us that from the occurrence of certain sacks in certain positions we can predict the occurrence of certain others in certain other positions, but it never tells us the structural relations among the contents of the sacks. The sample inference about the mortality of some angels cannot be made without "looking inside the sacks". It depends not only on the way the premises are related via sentential connectives; it depends also on the common occurrence of the word "angels" in the two premises, and, even more importantly, it depends on the relations between the logical words "some" and "all".

The proper context for a logical analysis of *some* and *all* is the theory of propositional functions. We have seen that a suitable algebraic interpretation of "proposition" is "element of a Boolean algebra" (possibly, but not necessarily, a free Boolean algebra). A *propositional function* (better: a proposition-valued function) is a function whose values are elements of a Boolean algebra. On the level of intuitive (non-algebraic) logic, if p_0 is the sentence (or proposition)

"6 is even", and if $p(x)$ is "x is even", then p is a propositional function that takes the value p_0 when $x = 6$.

To study propositional functions in suitable generality, fix an arbitrary non-empty set X, fix an arbitrary Boolean algebra $\mathbf{B}$, and, to begin with, consider the set $\mathbf{A}$ of all functions p from X to $\mathbf{B}$. (The algebra $\mathbf{B}$ is sometimes called the *value algebra* of $\mathbf{A}$.) If p is such a function (an element of $\mathbf{A}$), it is natural to define its complement to be the function $\neg p$ that is the pointwise complement of p:

$$(\neg p)(x) = \neg (p(x))$$

for all x; and if p and q are two such functions, it is equally natural to define their supremum and their infimum to be the functions $p \vee q$ and $p \wedge q$ that are the pointwise supremum and infimum of p and q:

$$(p \vee q)(x) = p(x) \vee q(x),$$

and

$$(p \wedge q)(x) = p(x) \wedge q(x).$$

(The operations on characteristic functions that were defined at the end of the section on lattices are special cases of this construction.) It is a routine and trivial task to verify that the set $\mathbf{A}$ (of functions) with the functional operations $\neg$, $\vee$, and $\wedge$ satisfies all the defining axioms of Boolean algebras: $\mathbf{A}$ itself is a Boolean algebra. It is sometimes referred to as the Xth *direct power* of $\mathbf{B}$ and is denoted by $\mathbf{B}^X$. The zero and the unit of $\mathbf{A}$ are (obviously) given by $0(x) = 0$ and $1(x) = 1$ for all x. As in any Boolean algebra, there is a natural partial ordering of $\mathbf{A} : p \leq q$ if and only if $p \wedge q = p$. Clearly, $p \leq q$ (in $\mathbf{A}$) just in case $p(x) \leq q(x)$ (in $\mathbf{B}$) for every x in X.

44. Finite functions

If that were all we could say about $\mathbf{A}$, there wouldn't have been much point in introducing it. The proper question is this: does the Boolean algebra $\mathbf{A}$ have any special structure that it derives from its construction out of X and $\mathbf{B}$? A partial answer is that in $\mathbf{A}$ there is a naturally distinguishable subalgebra $\mathbf{A}_0$, which is, in fact, naturally isomorphic to the value-algebra $\mathbf{B}$. The subalgebra $\mathbf{A}_0$ consists of the constant functions in A: the functions p such that $p(x) = p(y)$ for all

x and y in X. The unique value of such a function is an element of $\mathbf{B}$, and, clearly, there is a natural one-to-one correspondence between $\mathbf{A}_0$ and $\mathbf{B}$. The correspondence is the isomorphism that assigns to each constant function (in $\mathbf{A}_0$) its unique value (in $\mathbf{B}$); inversely, the isomorphism assigns to each element p_0 of $\mathbf{B}$ the function p in $\mathbf{A}_0$ defined by $p(x) = p_0$ for all x in X. There is no danger in regarding a constant function as identical with its value, and we shall usually do just that.

The function algebra $\mathbf{A}$ includes another, perhaps less obvious, subalgebra $\mathbf{A}_1$, whose definition, like that of the subalgebra $\mathbf{A}_0$ of constant functions, depends on the elements of $\mathbf{A}$ being not merely abstract objects in some postulationally defined Boolean algebra, but concrete functions on some set. The subalgebra $\mathbf{A}_1$ consists of the constant finite-valued functions in $\mathbf{A}$: the functions p such that ran(p) (the range of p) is a finite subset of $\mathbf{B}$. (Recall that ran(p) is, by definition, the set of all those elements q of $\mathbf{B}$ for which there exists an x in X with $p(x) = q$. A function in $\mathbf{A}$ is a constant function if and only if its range is a singleton — that is, its range contains exactly one element.) A moment's meditation is enough to prove that the set $\mathbf{A}_1$ is indeed a subalgebra of $\mathbf{A}$ (that is, $\mathbf{A}_1$ is closed under $\neg$, $\vee$, and $\wedge$); for example, if p and q are in $\mathbf{A}_1$, then ran $(p \vee q)$ is included in the finite set of elements of the form $r \vee s$, where r is in ran(p) and s is in ran(q). Clearly $\mathbf{A}_0 \subset \mathbf{A}_1$.

Intuitively, a proposition says something; a constant propositional function says the same thing about everything, and a finite-valued propositional function says only a finite number of different things about everything. Observe that if X is a singleton, then $\mathbf{A}$ is the same as $\mathbf{A}_0$; if either X is finite or $\mathbf{B}$ is finite, then $\mathbf{A}$ is the same as $\mathbf{A}_1$. If $\mathbf{B} = \mathbf{2}$, then $\mathbf{A}$ is the set of all characteristic functions on X, so that in that case $\mathbf{A}$ is naturally isomorphic to the field of all subsets of X. In this special case, $\mathbf{A}_0$ has just the two elements 0 and 1, and $\mathbf{A}_1 = \mathbf{A}$.

If $p \in \mathbf{A}_1$, then it makes sense to form the supremum (join) of all the (finitely many) values of p. If all the different values (in $\mathbf{B}$) that the element p (of $\mathbf{A}_1$) attains are $p_1, \dots, p_n$, then the supremum of the values of p is

$$\bigvee \operatorname{ran}(p) = \bigvee_{i=1}^{n} p_i = p_1 \vee \cdots \vee p_n.$$

According to this definition, $\bigvee \operatorname{ran}(p)$ is an element of $\mathbf{B}$. It is usually convenient to regard it as an element of $\mathbf{A}_0$ instead. To avoid all possible danger of confusion at this point, we introduce a neutral symbol: we propose to write Qp (or sometimes $Q(p)$) for the function (element of $\mathbf{A}$) defined for all x in

X by

$$Qp(x) = \bigvee \operatorname{ran}(p).$$

Observe that $Qp \in \mathbf{A}_0$, and therefore $Qp \in \mathbf{A}_1$.

Consider the mapping Q from $\mathbf{A}_1$ into $\mathbf{A}_1$ (and onto its subalgebra $\mathbf{A}_0$) that sends each p in $\mathbf{A}_1$ to the function Qp. What are its properties? How is it related to the Boolean structure of the Boolean algebra (namely $\mathbf{A}_1$) on which it is defined? Some of the answers are easy. Thus, for instance, it is clear that

$$Q0 = 0$$

(Q is *normalized*). It is also clear that

$$p \leq Qp$$

for all p in $\mathbf{A}_1$ (Q is *increasing*); indeed, this just says that each term that enters into a supremum is dominated by that supremum. In particular, $Q1 = 1$. The questions and the answers become a little more interesting when we ask about the relation of Q to $\vee$. The answer is the simplest possible:

$$Q(p \vee q) = Q(p) \vee Q(q)$$

for all p and q in $\mathbf{A}_1$ (Q is *distributive* over $\vee$). Another useful property of Q is expressed by the equation

$$Q(Qp) = Qp$$

for all p (Q is *idempotent*). Since $Qp \in \mathbf{A}_0$ for all p in $\mathbf{A}_1$, this says that the result of applying Q to a constant function is just that constant function, or, equivalently, that the supremum of the range of a constant function is its unique value.

The relation of Q and complementation is expressed by the equation

$$Q(\neg Qp) = \neg Qp;$$

the proof is the same as it was for idempotence. (The point is that, since Qp is in $\mathbf{A}_0$, so is $\neg Qp$.) The relation of Q and $\wedge$ is expressed by the equation

$$Q(p \wedge Qq) = Qp \wedge Qq.$$

The proof this time depends on the distributive law. Suppose that the range of p is $\{p_1, \ldots, p_n\}$. Since Qq is a constant function with range $\{Qq\}$ (recall that

Qq denotes both the constant function and its value), the range of $p \wedge Qq$ is the set

$$\{p_1 \wedge Qq, \ldots, p_n \wedge Qq\}.$$

Therefore

$$Q(p \wedge Qq) = \bigvee_{i=1}^{n}(p_i \wedge Qq) = \left(\bigvee_{i=1}^{n} p_i\right) \wedge Qq = Qp \wedge Qq.$$

The equation just derived essentially expresses the modularity of the operation Q over $\wedge$. Indeed, it says that

$$Q(p \wedge r) = Qp \wedge r,$$

provided r has the form Qq. The assertion that Q is modular over $\wedge$ is the same identity but with the superficially different proviso that r is fixed under Q. The difference is only apparent. If $r = Qq$, then

$$Qr = QQq = Qq = r,$$

so that r is fixed under Q; if $r = Qr$, then

$$r = Qq \text{ with } q = r.$$

In other words, for an idempotent operation the range is the same as its set of fixed elements.

As long as we are digressing, we might mention that the other similar derived property of Q, the identity $Q(\neg Qp) = \neg Qp$, is a kind of commutative law — it is a restricted kind of commutative law in the same sense in which the modular law is a restricted kind of distributive law. That is: to say that Q commutes with complementation would be to say that

$$Q \neg p = \neg Qp$$

for all p. The statement that this holds, not for all p but only for the p's that are fixed under Q (equivalently, for the p's in the range of Q), is exactly the one that was proved for Q above.

45. Functional monadic algebras

With this much technical machinery available, we can now generalize our study of propositional functions. Suppose again that X is a non-empty set and that

B is a Boolean algebra, and consider the set $\mathbf{B}^X$ of all functions from X to **B**. As we have seen, $\mathbf{B}^X$ is a Boolean algebra. For our purposes there is a natural way to enlarge the set $\mathbf{A}_1$ of all finite-valued functions considered before. What was essential about a function p in $\mathbf{A}_1$ was that its range had a supremum in **B**, and that condition makes sense whether the range of p is finite or not. We have now reached the goal of our discussion of functions with values in a Boolean algebra. What we shall be interested in from now on is Boolean subalgebras **A** of $\mathbf{B}^X$ satisfying two conditions:

(i) **A** contains every constant function,

and

(ii) **A** is included in the set of functions whose range has a supremum (in **B**).

(Note: the set **C** of functions whose range has a supremum is not necessarily a subalgebra of $\mathbf{B}^X$ — it is not necessarily closed under the formation of complements. If a Boolean subalgebra **A** of $\mathbf{B}^X$ is included in **C**, then for each p in **A** it must be true that both ran(p) and ran($\neg p$) have a supremum.) If **A** is a subalgebra of $\mathbf{B}^X$ satisfying these two conditions and if $p \in \mathbf{A}$, then we shall consider, as before, the function Qp, the constant function whose single value is $\bigvee \operatorname{ran}(p)$. The two conditions insure that Qp is always a well-defined function in **A**, so that Q is a mapping from **A** into **A** (and onto the algebra of constant functions). In other words, **A** is closed under the (unary) operation Q. The algebra consisting of the set **A** with the Boolean operations inherited from $\mathbf{B}^X$ and the operation Q will be called a *functional monadic algebra* (or, to give it its full title, an **A**-valued functional monadic algebra with domain X). The reason for the word "monadic" is that the concept of a monadic algebra (to be defined in appropriate generality below) is a special case of the concept of a polyadic algebra; the special case is characterized by the superimposition on the Boolean structure of exactly *one* additional operator.

Here is a simple but important example of a functional monadic algebra. Take **B** to be **2** and **A** to be $\mathbf{2}^X$. Conditions (i) and (ii) hold trivially for **A**. A moment's reflection suffices to clarify what the operation Q is. If p is the zero element of **A** (the function that is constantly 0), then the supremum of ran(p) is 0, and therefore $Qp = 0$. If p is a non-zero element of **A**, then $p(x) = 1$ for some point x of X. It follows that 1 is in ran(p), and hence that the supremum of ran(p) is 1. Thus, $Qp = 1$. In sum, $Qp = 0$ when $p = 0$, and $Qp = 1$ when $p \neq 0$. As was pointed out before, **A** can also be viewed as the collection of all subsets of X. From this perspective, Q is the operation on **A** that maps the empty set to itself and all other sets to X.

The algebraic properties of Q are as easy to derive in the general case of an arbitrary functional monadic algebra as they were in the case of the algebra of functions with finite ranges. We explicitly mention three of them: Q is normalized (which means that $Q0 = 0$), increasing (which means that $p \leq Qp$), and modular over $\wedge$ (which means that $Q(p \wedge Qq) = Qp \wedge Qq$). The proofs are the same as in the finite case, except, of course, that the distributive law needed to prove that Q is modular over $\wedge$ in the infinite case is the infinite distributive law.

46. Functional quantifiers

We ordinarily think of every statement (sentence, proposition) as being necessarily either "true" or "false" in some absolute sense. This makes it unintuitive and difficult to consider an example of a proposition that is neither being asserted nor denied, but just being said. If someone says "7 is even", you believe it, or you don't; the statement is true or it is false. In either case the situation is algebraically (Boolean algebraically) trivial. In order to have a non-trivial algebra (one that is different from the Boolean algebra **2**) in the role of the value algebra **B** that enters in the theory of propositional functions, let X be a non-empty set, let **B** be the Boolean algebra of all subsets of X, and consider the functional algebra $\mathbf{B}^X$. In this situation, if $x \in X$ and $p \in \mathbf{B}^X$, then $p(x)$ is a subset of X.

The Boolean algebra $\mathbf{B}^X$ is naturally isomorphic to the Boolean algebra of all subsets of the set $X \times X$ (the set of ordered pairs of elements of X) via the isomorphism that assigns to each p in $\mathbf{B}^X$ the set

$$\{(x, y) : y \in p(x)\}.$$

The range of an element p in $\mathbf{B}^X$ is a collection of subsets of X, and the supremum of this range (in **B**) is the union of this collection. Therefore Qp is the constant function whose value at each x in X is the union of the sets in the range of p. It follows that the set corresponding to Qp under the isomorphism is

$$\{(x, y) : \text{there exists a } z \text{ such that } y \in p(z)\},$$

that is, it is the set of pairs (of elements in X) whose second coordinate is in the union of $\operatorname{ran}(p)$.

It is frequently helpful to visualize the example in the preceding paragraph geometrically. If X is, say, the real line, then $\mathbf{B}^X$ is naturally isomorphic to the algebra of all subsets of the Cartesian plane, and the set that corresponds to Qp under this isomorphism is the union of all horizontal lines that pass through some point of the set corresponding to p. For instance, if p is the function that maps the number 1 to the set $\{2,4,6\}$, the number 2 to the set $\{3,5\}$, and every other real number to the empty set, then Qp is the constant function that maps every real number to the set $\{2,3,4,5,6\}$. Consequently, the isomorphism assigns to p the set

$$\{(1,2),(1,4),(1,6),(2,3),(2,5)\},$$

and to Qp the set of all points in the plane whose second coordinate is one of 2, 3, 4, 5, or 6 — and this is just the union of the horizontal lines through the points $(1,2)$, $(1,4)$, $(1,6)$, $(2,3)$, and $(2,5)$.

In view of the isomorphism between $\mathbf{B}^X$ and the Boolean algebra of subsets of $X \times X$, it makes sense to associate (temporarily) with a propositional function p in $\mathbf{B}^X$ the statement "y belongs $p(x)$". It follows from what we wrote above that the statement associated with Qp is "there exists an element z such that y belongs to $p(z)$". This indicates that the role of Q is that of an *existential quantifier*.

The algebraic study of propositional functions is related to their functional study (as pursued till now) the same way as the algebraic study of Boolean algebras is related to the study of fields of sets. The way to learn which properties of Q should be selected to play the role of axioms, so that other properties may be derived from them, is to experiment. Humanity has performed the necessary experiments, and we are ready to report the results: the three simple properties of Q already mentioned (Q is normalized, increasing, and modular) are sufficient to imply all others. We proceed to formulate the appropriate abstract definitions, and, at the same time, we change the notation (replace Q by $\exists$) so as to conform to standard usage and have a permanent reminder of what the subject is all about.

A *quantifier* (properly speaking, an existential quantifier) is a mapping $\exists$ of a Boolean algebra into itself that is normalized, increasing, and modular over $\wedge$. The concept of a *universal* quantifier is defined by the dualization

$$\forall p = \neg(\exists \neg p),$$

and therefore has the dual properties of the existential quantifier:

$$\forall 1 = 1,$$

$$p \geq \forall p,$$

$$\forall(p \vee \forall q) = \forall p \vee \forall q.$$

(Note the intuitive content of the equation defining the universal quantifier: something is always true if it's never false, that is, not ever not true.) The perfect duality between $\forall$ and $\exists$ justifies the asymmetric treatment in what follows; we shall study $\exists$ alone and content ourselves with an occasional comment on the behavior of $\forall$.

A word about our notation is in order. Because we are studying propositional functions of a single variable, it is not necessary to call attention notationally to the variable that is being quantified. That is why it suffices to use a symbol such as $\exists$ without any reference to a variable. If we were studying propositional functions of two variables, say x and y, then it would be necessary to indicate which variable was being quantified. There would, in fact, be two distinct existential quantifiers that might be denoted by $\exists x$ and $\exists y$.

47. Properties of quantifiers

Many of the properties of quantifiers in functional monadic algebras can actually be derived from the abstract definition of a quantifier. Throughout the following discussion we assume that **A** is a Boolean algebra and that $\exists$ is a quantifier on A.

Observe first that $\exists 1 = 1$ (since $1 \leq \exists 1$). Because

$$\exists(\exists q) = \exists(1 \wedge \exists q) = \exists 1 \wedge \exists q = 1 \wedge \exists q = \exists q,$$

it follows that $\exists$ is idempotent. As we have already seen, from idempotence we can conclude that the range of $\exists$ is equal to the set of elements fixed under $\exists$.

It is true that every (existential) quantifier is distributive over $\vee$, but the proof is slightly tricky. We proceed via some lemmas, as follows.

(i) If $p \leq \exists q$, then $\exists p \leq \exists q$.

Proof. By assumption $p \wedge \exists q = p$; it follows that

$$\exists p = \exists(p \wedge \exists q) = \exists p \wedge \exists q \leq \exists q.$$

(ii) If $p \leq q$, then $\exists p \leq \exists q$ ($\exists$ is monotone).

Proof. Note that $p \leq q \leq \exists q$ and apply (i).

(iii) $\exists(\neg \exists p) = \neg \exists p$.

Proof. Since $\neg \exists p \wedge \exists p = 0$, it follows from normality and modularity that

$$0 = \exists 0 = \exists(\neg \exists p \wedge \exists p) = \exists(\neg \exists p) \wedge \exists p,$$

so that

$$\exists(\neg \exists p) \leq \neg \exists p.$$

The reverse inequality is a consequence of the increasing character of $\exists$.

(iv) The range of $\exists$ is a Boolean subalgebra of **A**.

Proof. If p and q are in $\exists(\mathbf{A})$ (the range of $\exists$), then $p = \exists p$ and $q = \exists q$ (by idempotence), and therefore

$$p \wedge q = \exists p \wedge \exists q = \exists(p \wedge \exists q) \in \exists(A)$$

(by modularity). If p is in $\exists(\mathbf{A})$, then $p = \exists p$ and therefore (by (iii))

$$\neg p = \neg \exists p = \exists(\neg \exists p) \in \exists(A).$$

Here, finally, is distributivity: $\exists(p \vee q) = \exists p \vee \exists q$.

Proof. Since $p \leq p \vee q \quad \text{and} \quad q \leq p \vee q$, we have

$$\exists p \leq \exists(p \vee q) \quad \text{and} \quad \exists q \leq \exists(p \vee q)$$

(by (ii)), and therefore

$$\exists p \vee \exists q \leq \exists(p \vee q).$$

To prove the reverse inequality, observe first that both $\exists p$ and $\exists q$ belong to $\exists(\mathbf{A})$; therefore (by (iv)) $\exists p \vee \exists q$ belongs to $\exists(\mathbf{A})$. It follows that

$$\exists(\exists p \vee \exists q) = \exists p \vee \exists q.$$

Since

$$p \leq \exists p \vee \exists q \quad \text{and} \quad q \leq \exists p \vee \exists q$$

($\exists$ is increasing), we have $p \vee q \leq \exists p \vee \exists q$, and therefore (by (ii))

$$\exists(p \vee q) \leq \exists(\exists p \vee \exists q) = \exists p \vee \exists q.$$

The proof is complete.

The dual monotony law,

$$\text{if } p \le q, \text{ then } \forall p \le \forall q,$$

and the dual distributive law,

$$\forall(p \wedge q) = \forall p \wedge \forall q,$$

follow by duality.

It is sometimes useful to know the relation between quantification and subtraction, and the relation between quantification and Boolean addition. The result and its proof are simple. Assertion:

$$\exists p - \exists q \le \exists(p - q)$$

and

$$\exists p + \exists q \le \exists(p + q).$$

Proof. Since $p \vee q = (p - q) \vee q$, it follows from distributivity that

$$\exists p \vee \exists q = \exists(p \vee q) = \exists((p - q) \vee q) = \exists(p - q) \vee \exists q;$$

form the infimum of both sides with $\neg \exists q$ and conclude that

$$\exists p - \exists q = \exists(p - q) - \exists q \le \exists(p - q).$$

The result for Boolean addition follows from two applications of the result for subtraction and from the distributive law for $\exists$ over suprema.

48. Monadic algebras

A *monadic algebra* is a Boolean algebra **A** together with an existential quantifier $\exists$ on **A**. Every functional monadic algebra is (as the name implies) an example of a monadic algebra. Here are two other easy examples of monadic algebras that are at least prima facie different from the functional examples.

(i) Any Boolean algebra with the identity mapping as the quantifier; this quantifier will be called *discrete*.

(ii) Any Boolean algebra with the mapping $\exists$ defined by

$$\exists 0 = 0 \quad \text{and} \quad \exists p = 1 \quad \text{for all } p \ne 0$$

as the quantifier; this quantifier will be called *simple*.

The elementary algebraic theory of monadic algebras is similar to that of every other algebraic system, and, consequently, it is rather a routine matter. Thus, for example, a subset **B** of a monadic algebra **A** is a *monadic subalgebra* of **A** if it is a Boolean subalgebra of **A** and if it is a monadic algebra with respect to the quantifier on **A**. Equivalently, a Boolean subalgebra **B** of **A** is a monadic subalgebra of **A** if and only if $\exists p \in \mathbf{B}$ whenever $p \in \mathbf{B}$. The central concept is, as usual, that of a homomorphism; *a monadic homomorphism* is a mapping f from one monadic algebra into another, such that f is a Boolean homomorphism and

$$f(\exists p) = \exists f(p)$$

for all p. Associated with every homomorphism is its kernel $\{p : f(p) = 0\}$. The kernel of a monadic homomorphism is a *monadic ideal*; that is, it is a Boolean ideal **I** in **A** such that $\exists p \in \mathbf{I}$ whenever $p \in \mathbf{I}$. Similarly, a monadic filter in **A** is a Boolean filter **F** in **A** such that $\forall p \in \mathbf{F}$ whenever $p \in \mathbf{F}$. A monadic congruence relation in **A** is a Boolean congruence relation $\equiv$ in **A** such that $\exists p \equiv \exists q$ whenever $p \equiv q$. The same relation exists between monadic ideals, filters, and congruence relations as exists between their Boolean counterparts. The adjective "monadic" will be used with "subalgebra", "homomorphism", etc., whenever it is advisable to emphasize the distinction from other kinds of subalgebras, homomorphisms, etc. — for example, from the plain Boolean kind. Usually, however, the adjective will be omitted and the context will unambiguously indicate what is meant.

The definition of monadic quotient algebras and the consequent homomorphism theorem (every proper ideal is a kernel) work as usual. If **A** is a monadic algebra and **I** is a monadic ideal (or **F** is a monadic filter) in **A**, form the Boolean quotient algebra $\mathbf{B} = \mathbf{A}/\mathbf{I}$ (or $\mathbf{A}/\mathbf{F}$), and consider the canonical Boolean homomorphism f from **A** onto **B** that takes every element p to its equivalence class $[p]$ modulo **I**. There is a unique natural way of converting **B** into a monadic algebra so that f becomes a monadic homomorphism (with kernel **I**, of course): define $\exists[p]$ (in **B**) to be $[\exists p]$ (where $\exists p$ is an element of **A**). To show that this existential quantifier is well defined in **B**, suppose that p_1 and p_2 are in **A** and that $[p_1] = [p_2]$. Then $p_1 + p_2$ is in **I**, and therefore so is $\exists(p_1 + p_2)$ (since **I** is a monadic ideal). It follows that $\exists p_1 + \exists p_2$ is in **I** (because $\exists p_1 + \exists p_2 \leq \exists(p_1 + p_2)$), whence

$$[\exists p_1] = [\exists p_2].$$

A monadic algebra is *simple* if $\{0\}$ is the unique proper ideal in it. A monadic ideal is *maximal* if it is a proper ideal that is not a proper subset of any other proper ideal. The connection between maximal ideals and simple algebras is an elementary part of universal algebra: the kernel of a homomorphism is a maximal ideal if and only if the range of the homomorphism is a simple algebra.

The connection between simplicity and quantifiers is simple: a monadic algebra is simple if and only if it is non-trivial and its quantifier is simple.

Proof. If $\mathbf{A}$ is simple, then $\{0\}$ is a proper ideal, so there is certainly a non-zero element in $\mathbf{A}$. If p is any non-zero element of $\mathbf{A}$, write $\mathbf{I} = \{q : q \leq \exists p\}$. Since, clearly, $\mathbf{I}$ is a non-trivial monadic ideal, it follows that $\mathbf{I} = \mathbf{A}$, and hence, in particular, that $1 \in \mathbf{I}$. This implies that $\exists p = 1$ whenever $p \neq 0$, so the existential quantifier of $\mathbf{A}$ is simple. Suppose, conversely, that $\mathbf{A}$ is non-trivial and that $\exists p = 1$ whenever $p \neq 0$. Certainly $\{0\}$ is a proper ideal of $\mathbf{A}$. If $\mathbf{I}$ is a non-trivial monadic ideal in $\mathbf{A}$, then it must contain a non-zero element p. Therefore $\exists p$ is in $\mathbf{I}$ (because $\mathbf{I}$ is monadic), so that 1 is in $\mathbf{I}$ (because $\exists$ is simple). Hence $\mathbf{I} = \mathbf{A}$. In other words, every non-trivial ideal in $\mathbf{A}$ is improper; this proves that $\mathbf{A}$ is simple.

It is obvious that a subalgebra of a non-trivial monadic algebra $\mathbf{A}$ with a simple quantifier is itself non-trivial and inherits the simple quantifier of $\mathbf{A}$. This observation and the characterization of simple algebrasin the preceding paragraph lead to the following conclusion.

Theorem. *Every subalgebra of a simple monadic algebra is simple.*

The only simple *Boolean* algebra is the two-element algebra $\mathbf{2}$. Since a monadic algebra is simple if and only if its quantifier is simple, and since (as we have seen) the quantifier of the functional monadic algebra $\mathbf{2}^X$ is simple, it follows that $\mathbf{2}^X$ is a simple monadic algebra whenever X is non-empty. The preceding theorem implies that every subalgebra of $\mathbf{2}^X$, that is, every $\mathbf{2}$-valued functional (monadic) algebra with a non-empty domain, is also simple. Up to isomorphisms they are the only simple algebras.

Theorem. *A monadic algebra is simple if and only if it is (isomorphic to) a* $\mathbf{2}$*-valued functional monadic algebra with a non-empty domain.*

Proof. It has already been shown that every $\mathbf{2}$-valued functional algebra with a non-empty domain is simple. The converse is just as easy to prove, but the proof makes use of a relatively deep fact, namely Stone's theorem on the representation of Boolean algebras.

If **A** is a simple monadic algebra, then **A** is, in particular, a non-trivial Boolean algebra, to which Stone's theorem (in the form that refers to a Boolean algebra of characteristic functions) is applicable. It follows that there exist (i) a non-empty set X, (ii) a Boolean subalgebra **B** of $\mathbf{2}^X$, and (iii) a Boolean isomorphism f from **A** onto **B**. If **B** is treated as a monadic subalgebra of $\mathbf{2}^X$, then it is certainly a **2**-valued functional algebra (whose quantifier is simple). It is straightforward to check that f must also preserve quantification. Indeed, f preserves 0 and 1 (it is a Boolean isomorphism), and the quantifiers in both **A** and **B** are simple, and therefore take on the value 1 at all non-zero elements. Conclusion: f is a monadic isomorphism between the monadic algebras **A** and **B**.

In the preceding proof we started with a monadic algebra **A**, treated it as a Boolean algebra, and then looked at it again as a monadic algebra. This procedure is applied frequently in the study of monadic algebras. For example, there is a close connection between the Boolean ideals and the monadic ideals of **A**. For every Boolean ideal **I** of **A**, let $\mathbf{I}^*$ be the set of all those p in **A** for which $\exists p$ is in **I**. It is a simple matter to check that $\mathbf{I}^*$ is a monadic ideal. Indeed, if p and q are in $\mathbf{I}^*$, then $\exists p$ and $\exists q$ are in **I**, by definition, and therefore $\exists p \vee \exists q$ is in **I**. It follows from the distributive law that $\exists(p \vee q)$ is in **I**, so $p \vee q$ is in $\mathbf{I}^*$. If p is in **A** and q is in $\mathbf{I}^*$, then $\exists q$ is in **I**, and hence so is $\exists p \wedge \exists q$. From the modular law it follows that $\exists(p \wedge \exists q)$ is in **I**, and therefore so is $\exists(p \wedge q)$ (because $\exists q$ is increasing). Thus, $p \wedge q$ is in $\mathbf{I}^*$. If p is in $\mathbf{I}^*$, then $\exists p$ is in **I**. Idempotence implies that $\exists\exists p$ is in **I**, so $\exists p$ is in $\mathbf{I}^*$. This proves that $\mathbf{I}^*$ is a monadic ideal. If the correspondence that takes each Boolean ideal **I** to $\mathbf{I}^*$ leaves **I** invariant, that is, if $\mathbf{I} = \mathbf{I}^*$, then **I** is a monadic ideal. The converse is trivial: monadic ideals are fixed under the correspondence. It is clear from the definition of $\mathbf{I}^*$ that the correspondence preserves inclusion: if $\mathbf{I} \subset \mathbf{J}$, then $\mathbf{I}^* \subset \mathbf{J}^*$.

If **I** is a maximal Boolean ideal of **A**, then $\mathbf{I}^*$ is a maximal monadic ideal. To see this, suppose that **M** is a monadic ideal properly extending $\mathbf{I}^*$. Then there is an element p in **M** that is not in $\mathbf{I}^*$. Because **M** is a monadic ideal, it contains the element $\exists p$. Obviously, $\exists p$ is not in **I** (otherwise p would be in $\mathbf{I}^*$). Since **I** is a maximal Boolean ideal, the element $\neg \exists p$ must be in **I**. But this means that $\exists \neg \exists p$ is in **I**. Therefore $\neg \exists p$ is in $\mathbf{I}^*$ and hence also in **M**. Since **M** contains both an element and its complement, it is improper.

In sum, the correspondence that takes each Boolean ideal $\mathbf{I}$ to $\mathbf{I}^*$ maps the set of Boolean ideals of $\mathbf{A}$ onto the set of monadic ideals of $\mathbf{A}$, preserves inclusion, and takes maximal ideals to maximal ideals. The ideals left invariant by the correspondence are precisely the monadic ideals.

The maximal ideal theorem for Boolean algebras generalizes to monadic algebras. The proof for monadic algebras can be carried out by a monadic imitation of the Boolean proof, but it also follows directly from the Boolean theorem with the help of the preceding remarks.

Maximal ideal theorem for monadic algebras. *Every proper ideal in a monadic algebra is included in some maximal ideal.*

Proof. If $\mathbf{I}_0$ is a proper monadic ideal of a monadic algebra $\mathbf{A}$, then $\mathbf{I}_0$ is a proper Boolean ideal, and hence can be extended to a maximal Boolean ideal $\mathbf{I}$. By the preceding remarks, the set $\mathbf{I}^*$ is a maximal monadic ideal and $\mathbf{I}_0^* \subset \mathbf{I}^*$. Since $\mathbf{I}_0^* = \mathbf{I}_0$ by assumption, $\mathbf{I}^*$ is a maximal monadic ideal extending $\mathbf{I}_0$.

The existence theorem for monadic algebras follows from the maximal ideal theorem just as for Boolean algebras.

Existence theorem for monadic algebras. *For every non-zero element p_0 of every monadic algebra $\mathbf{A}$, there is a homomorphism f from $\mathbf{A}$ onto a simple monadic algebra such that $f(p_0) \neq 0$.*

Proof. Because of the relationship between maximal ideals and simple monadic algebras, the existence theorem can be rephrased as follows: there exists a maximal monadic ideal $\mathbf{I}$ in $\mathbf{A}$ such that $p_0 \notin \mathbf{I}$. To get one, apply the maximal ideal theorem to the ideal $\mathbf{I}_0$ of all those elements p of $\mathbf{A}$ for which $p \leq \neg \exists p_0$. If p_0 were in $\mathbf{I}$, then the element $p_0 \vee \neg \exists p_0$ would be in $\mathbf{I}$, and hence so also would the element $\exists(p_0 \vee \neg \exists p_0)$ (since $\mathbf{I}$ is closed under quantification). But the latter element coincides with $\exists p_0 \vee \neg \exists p_0$. Therefore $\mathbf{I}$ would contain 1, and hence would be improper. Thus, $p_0 \notin \mathbf{I}$. The proof is complete.

Under what conditions on a monadic algebra $\mathbf{A}$ is it true that whenever an element p of $\mathbf{A}$ is mapped to 0 by every homomorphism from $\mathbf{A}$ into a simple algebra, then $p = 0$? Since every monadic subalgebra of a simple monadic algebra is also simple, the difference between "into" and "onto" is not essential here. Because of the correspondence between homomorphisms with simple ranges and maximal ideals, the question could also be put this way: under what conditions on a monadic algebra **A** is it true that whenever an element p of **A** belongs to all maximal ideals, then $p = 0$? In analogy with other parts of algebra, it is natural to say that a monadic algebra is *semisimple* if the intersection of all maximal ideals in **A** is $\{0\}$. The question now becomes: which monadic algebras are semisimple? The answer — an immediate consequence of the existence theorem — is quite satisfying.

Semisimplicity theorem. *Every monadic algebra is semisimple.*

It should be remarked that the theorem is not universally true of all algebraic systems. There are examples of groups and examples of Boolean algebras with operators (generalizations of monadic algebras) that are not semisimple.

Since monadic algebras constitute a generalization of Boolean algebras, the theorem asserts, in particular, that every Boolean algebra is semisimple. This consequence of the theorem is well known: it is an immediate consequence of Stone's theorem, and, for us, it occurred as the most important step in the proof of that theorem, namely in the step where it is shown that the function f is one-to-one.

The theory of monadic algebras is much richer than has been indicated so far, but we shall not now enter into its further development. Two aspects of the theory deserve, however, at least passing mention: the representation of monadic algebras and the concept of freeness.

Another way of saying that a monadic algebra **A** is semisimple is to say that it is isomorphic to a subalgebra of a direct (Cartesian) product of simple monadic algebras (the simple algebras being the quotients $\mathbf{A}/\mathbf{I}$, where **I** ranges over the maximal ideals of **A**). We have seen that the simple monadic algebras coincide with the **2**-valued functional monadic algebras. Thus, an equivalent formulation of the previous theorem is that every monadic algebra is isomorphic to a subalgebra of a direct product of **2**-valued functional algebras. This is one example of a representation theorem for monadic algebras.

There is a somewhat stronger and deeper representation theorem (which we shall not prove): every monadic algebra is isomorphic to a functional monadic algebra. Why is this stronger? A functional monadic algebra is a subalgebra of $\mathbf{B}^X$ for some Boolean algebra **B** and some set X. By Stone's theorem, the algebra **B** is isomorphic to a subalgebra of the Boolean algebra $\mathbf{2}^Y$ for some set Y. Now it is not difficult to show that the functional monadic algebra $(\mathbf{2}^Y)^X$ (with domain X and *Boolean* value algebra $\mathbf{2}^Y$) is isomorphic to the monadic algebra $(\mathbf{2}^X)^Y$ (the direct product of copies of the single *monadic* algebra $\mathbf{2}^X$, one copy for each element of Y). Thus, $\mathbf{B}^X$ is isomorphic to a subalgebra of a direct product of $\mathbf{2}^X$. It follows that if every monadic algebra is isomorphic to a functional monadic algebra, then every monadic algebra is isomorphic to a subalgebra of a direct product of (copies of) a *single* **2**-valued functional monadic algebra.

49. Free monadic algebras

The concept of a free monadic algebra is easy to define, and the construction of free monadic algebras is similar to the construction of free Boolean algebras. The development of the propositional calculus can be imitated in almost every detail; indeed, from the symbolic point of view, the only difference is the appearance in the monadic theory of one new letter, namely $\exists$.

Some curious questions arise, and some of them have amusing answers. For instance, it is not always true, even for relatively simple algebraic systems, that a finitely generated algebra is finite. Example: the additive group of integers is generated by the single element 1. There is also a simple generalization of a monadic algebra, called a dyadic algebra, where finitely generated algebras may be infinite. It is easy to prove, however, that the free *Boolean* algebra on a finite number of generators is finite, and in fact it is easy to determine exactly how many elements it has; for monadic algebras the question is considerably trickier.

The lowest cases are easy enough: the free monadic algebra on zero generators is the Boolean algebra **2** (with, of course, the simple quantifier), and the free monadic algebra on one generator has sixteen elements. To see this, consider the generator p and its complement $\neg p$. The monadic algebra they generate contains the four basic elements

$$\forall p, \quad p \wedge \exists \neg p, \quad \neg p \wedge \exists p, \quad \text{and} \quad \forall \neg p.$$

As the diagram on the next page indicates, these four elements are disjoint and their supremum is 1. (Indeed, the elements $p \wedge \exists \neg p$ and $\neg p \wedge \exists p$ are clearly disjoint. With the help of the Boolean distributive law, their supremum can be computed to be $\exists p \wedge \exists \neg p$:

$$\begin{aligned}
&(p \wedge \exists \neg p) \vee (\neg p \wedge \exists p) \\
&\quad = (p \vee \neg p) \wedge (p \vee \exists p) \wedge (\neg p \vee \exists \neg p) \wedge (\exists p \vee \exists \neg p) \\
&\quad = \exists p \wedge \exists \neg p \wedge (\exists p \vee \exists \neg p) = \exists p \wedge \exists \neg p.
\end{aligned}$$

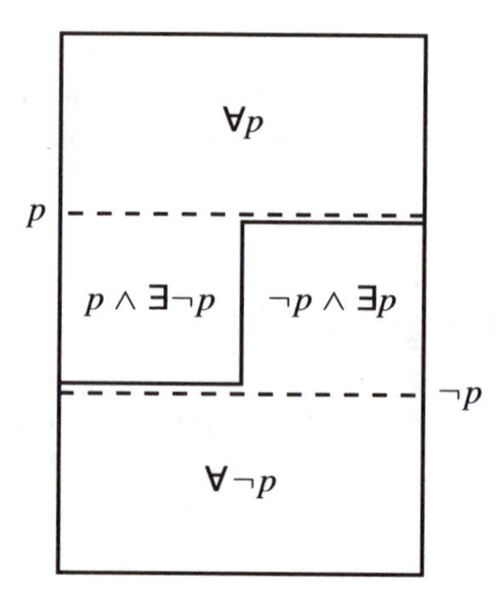

Since the elements $\exists p \wedge \exists \neg p$, $\neg \exists p$, and $\neg \exists \neg p$ are obviously disjoint and have 1 as their supremum, and since $\forall p$ coincides with $\neg \exists \neg p$ and $\forall \neg p$ with $\neg \exists p$, the assertion follows.) The existential quantifier leaves the elements $\forall \neg p$ and $\forall p$ fixed, and sends both $p \wedge \exists \neg p$ and $\neg p \wedge \exists p$ to the element $\exists p \wedge \exists \neg p$, by modularity. Therefore every element of the algebra is obtained by forming the supremum of some subset of the four basic elements. (In connection with the diagram, the quantifier applied to any element q yields the smallest horizontal strip that contains the basic elements of which q is the supremum.)

How many elements does the free monadic algebra on two generators have? Hyman Bass, who heard this question when he was a graduate student at the University of Chicago, became interested and a week later came up with the answer. He said that it's easy; the answer is obviously 4,294,967,296. When pressed, Bass confessed that he had the general answer; for n generators, the total number of elements is

$$2^{2^n \cdot 2^{2^n - 1}};$$

for $n = 2$ this becomes the "obvious" answer above.

50. Modal logics

Before examining in more detail the connections of monadic algebras with logic, we mention one more thing in passing, and that is the theory of *modal* logics. That theory is usually of most interest to philosophers. It is motivated by the feeling that logic ought to make precise philosophical words other than truth and falsehood — that concepts such as plausibility and possibility can be and deserve to be treated formally.

We don't know all the various proposals that have been made in this direction, but we do know one of them. If in addition to "p is true" ($\vdash p$) and "p is false" ($\vdash \neg p$), it is desired to discuss "p is possible", then one way to try to do that (within a Boolean algebra) is to introduce a "possibility operator" that converts each proposition p into a proposition Mp (for "maybe p"?), to endow M with suitable algebraic properties, and then to derive other properties from them. A reasonable list of properties for M is that it be normalized (if p is false, then so is "maybe p"), increasing (the proposition p entails the proposition "maybe p"), and modular over $\wedge$ ("maybe (p and maybe q)" is the same as "maybe p and maybe q"). From this point of view, M is nothing more or less than an existential quantifier, and this branch, at any rate, of the theory of modal logics is subsumed under (coincides with?) the theory of monadic logics.

51. Monadic logics

In the usual logical treatment of Boolean algebras and their generalizations, certain elements of the appropriate Boolean algebra are singled out and called "provable". From the algebraic point of view, the definition of provability in any particular case is irrelevant; what is important is the algebraic structure of the set of all provable elements. It is sometimes convenient, in the examination of that structure, to dualize, that is, to consider not provability but refutability. There is an obvious relation between the two concepts; clearly p should be called refutable if and only if $\neg p$ is provable.

Suppose now that **A** is a monadic algebra whose elements, for heuristic purposes, are thought of as propositions, or, rather, as propositional functions. What properties does it seem reasonable to demand of a subset **F** of **A** in order that its elements deserve to be called provable? Clearly, if p and q are provable, then $p \wedge q$ should also be provable, and if p is provable, then $p \vee q$ should be provable no matter what q may be. In other words, **F** should be, at least, a Boolean filter in **A**. That is not enough, however; **F** should also bear the proper relation to quantification. If, in other words, p is provable (and here it is essential that p be thought of as a propositional function, and not merely as a proposition), then $\forall p$ should also be provable. The requirement (satisfied by the set of provable elements in the usual logical algebras) converts **F** into a monadic filter.

The following definition is now adequately motivated: a *monadic logic* is a pair $(\mathbf{A}, \mathbf{F})$, where $\mathbf{A}$ is a monadic algebra and $\mathbf{F}$ is a monadic filter in $\mathbf{A}$. The elements p of $\mathbf{F}$ are the *provable* elements of the logic; if $\neg p \in \mathbf{F}$, then p is *refutable*. (In analogy with our definition of a Boolean logic, we could define a monadic logic to be a pair $(\mathbf{A}, \mathbf{F})$, where $\mathbf{A}$ is a "pre-monadic algebra" — a pre-Boolean algebra with an additional unary operation — and $\mathbf{F}$ is a congruence relation on $\mathbf{A}$ such that $\mathbf{A}/\mathbf{F}$ is a monadic algebra. However, as for Boolean logics, here too the apparently more general theory can be recaptured from the special case by passing from $\mathbf{A}$ to $\mathbf{A}/\mathbf{F}$; see the remark preceding the deduction theorem for Boolean logics.)

The questions of consistency and completeness play as important a role in monadic logic as they do in the propositional calculus. They are closely connected to representation theory for monadic algebra, just as the analogous questions for the propositional calculus are closely connected to representation theory for Boolean algebra. To formulate the relevant theorems it is necessary to specify "concrete" models; the semisimplicity theorem hints at what these models should be. A *model* is, by definition, a monadic logic $(\mathbf{A}, \mathbf{F})$, where $\mathbf{A}$ is a $\mathbf{2}$-valued functional monadic algebra with a non-empty domain and $\mathbf{F}$ is the trivial filter $\{1\}$. It is natural to consider $\mathbf{2}$-valued functional monadic algebras because (as was pointed out before) they can be viewed as fields of sets, enriched with the simple monadic quantifier. Note that since a $\mathbf{2}$-valued functional monadic algebra with a non-empty domain is simple, $\mathbf{F}$ could only be $\{1\}$ or $\mathbf{A}$, and the latter choice is obviously uninteresting.

An *interpretation* of a monadic logic $(\mathbf{A}, \mathbf{F})$ in a model $(\mathbf{B}, \{1\})$ is a monadic homomorphism f from $\mathbf{A}$ into $\mathbf{B}$ such that $f(p) = 1$ whenever $p \in \mathbf{F}$. A convenient way of expressing the condition is to say that every provable element is *true* in the interpretation. Equivalently, an interpretation of $(\mathbf{A}, \mathbf{F})$ in $(\mathbf{B}, \{1\})$ is a monadic homomorphism from $\mathbf{A}/\mathbf{F}$ into $\mathbf{B}$.

The logic $(\mathbf{A}, \mathbf{F})$ is said to be *semantically consistent* if there exists an interpretation of it. It is an easy consequence of the maximal ideal theorem for monadic algebras that $(\mathbf{A}, \mathbf{F})$ is consistent just in case the filter $\mathbf{F}$ is proper. Indeed, if $\mathbf{F}$ is improper, then it contains 0, and no homomorphism with a non-trivial range can map 0 to 1. On the other hand, if $\mathbf{F}$ is proper, then there is a maximal monadic filter $\mathbf{G}$ that includes it. The canonical homomorphism f maps $(\mathbf{A}, \mathbf{F})$ to $(\mathbf{A}/\mathbf{G}, \{1\})$; moreover, $(\mathbf{A}/\mathbf{G}, \{1\})$ is (isomorphic to) a model since $\mathbf{A}/\mathbf{G}$ is simple (because $\mathbf{G}$ is maximal). Therefore, f is an interpretation (at least up to isomorphisms).

An element p of $\mathbf{A}$ is called *valid* whenever it is true in every interpretation. By definition, every provable element is valid. There could conceivably be elements in $\mathbf{A}$ that are not provable but that are nevertheless valid. If there are no such elements, that is, if every valid element is provable, the logic is said to be *semantically complete*. This definition can also be expressed in dual form: a logic is semantically complete if every contravalid element is refutable. (The element p is *contravalid* if $f(p) = 0$ for every interpretation, that is, if p is *false* in every interpretation.) Elliptically but suggestively, semantic completeness can be described by saying that everything true is provable.

Semantic completeness demands of a logic $(\mathbf{A}, \mathbf{F})$ that the filter $\mathbf{F}$ be relatively large. If, in particular, $\mathbf{F}$ is very large, that is, $\mathbf{F} = \mathbf{A}$, then the logic is semantically complete simply because every element of $\mathbf{A}$ is provable. (The fact that in this case there are no interpretations is immaterial.) If $\mathbf{F} \neq \mathbf{A}$, then the quotient algebra $\mathbf{A}/\mathbf{F}$ may be formed, and the problem of deciding whether or not the logic $(\mathbf{A}, \mathbf{F})$ is semantically complete reduces to a question about the algebra $\mathbf{A}/\mathbf{F}$.

Since every interpretation of $(\mathbf{A}, \mathbf{F})$ in a model $(\mathbf{B}, \{1\})$ induces in a natural way a homomorphism from $\mathbf{A}/\mathbf{F}$ into $\mathbf{B}$, and since the only restriction on $\mathbf{B}$ is that it be simple, the question of semantic completeness, in its dual version, becomes the following one: under what conditions on the quotient algebra $\mathbf{A}/\mathbf{F}$ is it true that whenever an element p of $\mathbf{A}/\mathbf{F}$ is mapped to 0 by every homomorphism from $\mathbf{A}/\mathbf{F}$ into a simple algebra, then $p = 0$? As was pointed out before, another way of phrasing this question is: when is $\mathbf{A}/\mathbf{F}$ semisimple? The answer is given by the semisimplicity theorem: always! We have proved the following theorem.

Semantic completeness theorem for monadic logics. *Every monadic logic is semantically complete.*

52. Syllogisms

Much of classical logic was concerned with Aristotelean syllogistics, and even today there are mathematicians who are convinced that all their thinking can be formulated in syllogisms. That isn't so, but syllogisms are at least as interesting as the propositional calculus, and they are a much more traditional part of formal logic. The purpose of the immediate sequel is to show how the theory of syllogisms finds a simple and natural expression in the framework of monadic algebras. In what follows we shall be working with one fixed monadic algebra; in order to make sure that it has no special properties that might reduce a part of the theory to a triviality, it is sufficient to assume that the fixed algebra under consideration is a free monadic algebra on at least three generators.

Given the monadic algebra **A**, define four binary operations on **A** — that is, four functions from the set $\mathbf{A} \times \mathbf{A}$ of ordered pairs of elements in **A** to **A** — called A, E, I, and O, as follows:

$$\mathrm{A}(p,q) = \forall(p \Rightarrow q),$$

$$\mathrm{E}(p,q) = \forall(p \Rightarrow \neg q),$$

$$\mathrm{I}(p,q) = \exists(p \wedge q),$$

$$\mathrm{O}(p,q) = \exists(p \wedge \neg q).$$

A popular verbal rendering of these four expressions is based on interpreting the arguments p and q as properties (manhood, mortality, etc.) rather than as the corresponding propositional functions ("x is a man", "x is mortal", etc.), and reads, therefore as follows:

A: every p is q,

E: no p is q,

I: some p is q,

O: some p is not q.

The names (A, E, I, O) of the four expressions were fixed by the scholastics and are sacrosanct. They come from the root vowels of the Latin words

A F F I R M O and N E G O,

and are thus intended to serve as a reminder that A and I are affirmative, whereas E and O are negative. Also: A and E are universal, and I and O are particular.

Associated with the four operations A, E, I, O there is a large amount of taxonomic terminology. We do not know it all; we report here what we do know, partly just for fun, partly so that when you see it elsewhere, you will recognize it and not be slowed down and bewildered by it.

Since

$$\neg O(p,q) = \neg \exists(p \wedge \neg q) = \forall \neg (p \wedge \neg q) = \forall(\neg p \vee q) = A(p,q)$$

and

$$\neg E(p,q) = \neg \forall(\neg p \vee \neg q) = \exists \neg (\neg p \vee \neg q) = \exists(p \wedge q) = I(p,q),$$

each of O and A is called the *contradictory* of the other, and the same for E and I. These equations are also known as the *rules of opposition*. Since, trivially,

$$E(q,p) = \forall(\neg q \vee \neg p) = \forall(\neg p \vee \neg q) = E(p,q)$$

and

$$I(q,p) = \exists(q \wedge p) = \exists(p \wedge q) = I(p,q),$$

each of E and I is its own *converse*; these equations (that is, the assertions that E and I are symmetric in p and q) are the *rules of conversion*. Along the same lines, finally, are the equations

$$A(\neg q, \neg p) = \forall(\neg q \Rightarrow \neg p) = \forall(p \Rightarrow q) = A(p,q)$$

and

$$O(\neg q, \neg p) = \exists(\neg q \wedge p) = \exists(p \wedge \neg q) = O(p,q);$$

they are known as the rules of *contraposition*. Associated with these concepts is the *square of opposition*

A I
O E

(often

A O
I E

instead). Its columns are contradictories; its diagonals are called *contraries*. Here is a mnemonic for remembering these mnemonics: the four letters in the square of opposition are the four vowels in

A R I S T O T L E.

The principal problem of traditional logic is to classify all syllogisms. In order, first of all, to define syllogisms, we introduce the notation F*, whenever F is a binary operation on $\mathbf{A}$ (that is, whenever F is a mapping from $\mathbf{A} \times \mathbf{A}$ to $\mathbf{A}$) by writing

$$F^*(p, q) = F(q, p).$$

A *syllogism*, in terms of this notation, is an ordered triple

$$(F_1, F_2, F_3)$$

of functions such that each of F_1 and F_2 is either one of A, E, I, O, or else one of A^*, E^*, I^*, O^*, and F_3 is one of A, E, I, O. A syllogism is called *valid* if, for all elements p, q, and r, the element $F_3(p, r)$ belongs to the monadic filter generated by $F_1(q, r)$ and $F_2(p, q)$; in slightly different words,

$$F_1(q, r) \wedge F_2(p, q) \leq F_3(p, r)$$

for all p, q, and r (in ordinary language: $F_3(p, r)$ can be inferred from $F_1(q, r)$ and $F_2(p, q)$). Prima facie there are $8 \times 8 \times 4 = 256$ possible syllogisms; the classification problem is to select the valid ones from among them. Before discussing that problem, we report on the classical terminology associated with syllogisms.

For each binary operation, the first argument is called the *subject* and the second one the *predicate*. In a syllogism (F_1, F_2, F_3) the entries F_1 and F_2 are the *premises* and F_3 is the *conclusion*. The subject of the conclusion is the *minor term* of the syllogism, and the predicate of the conclusion is its *major term*. The subject of F_1, which is, of course, the same as the predicate of F_2, is called the *middle term* of the syllogism. The premise (namely F_2) that contains the minor term is the *minor premise*; the other premise (namely F_1) that contains the major term, is the *major premise*.

The *mood* of a syllogism is an ordered pair of the vowels A, E, I, O; the first term describes the major premise, and the second one the minor premise ("describes", not "is", because no asterisks appear in the mood). Example: the mood of (I, A^*, I) is (I, A). A syllogism belongs to one of four *figures*: the first has no asterisks; in the second, the major premise only has one; in the third, the minor premise only has one; and in the fourth, both premises have one.

Traditionally, valid syllogisms (and a small number of invalid ones) are associated with three-syllable mnemonic words whose vowels indicate the functions (A, E, I, or O) that occur in the major premise, the minor premise, and the conclusion, in that order. There are fifteen famous mnemonic words of this kind; here they are, classified according to the figures of the syllogisms they are intended to describe:

1. Barbara, Celarent, Carii, Ferio
2. Baroco, Camestres, Cesare, Festino
3. Bocardo, Datisi, Disamis, Feriso
4. Calemes, Dimatis, Fresison.

For example, Baroco describes the syllogism (A^*, O, O), Bocardo the syllogism (O, A^*, O), and Calemes the syllogism (A^*, E^*, E). The scholastics attached a certain mnemonic value to the consonants too, but it's perfectly safe not to know what they are.

A source of disagreement between traditional and modern logic is its treatment of the empty set (vacuous reasoning). Aristotle didn't believe in it; for him, for instance, $\exists(p \wedge q)$ is a consequence of $\forall(p \Rightarrow q)$. This situation accounts for the existence of four additional mnemonic words (intended to describe syllogisms then believed to be valid); namely

5. Darapti, Felapton
6. Bemalip, Fesapo.

The terminology described above used to be taken very seriously. As it was taught in, for instance, the early 1930's, the subject was almost exclusively syllogisms and their classification. Nowadays, much of the terminology is recognized as utterly pointless, since many of the distinctions it makes are based not on structure, but on notation alone. Thus, for instance, since $E = E^*$, there is no legitimate difference between Celarent and Cesare; a high-handed disregard of the commutative laws succeeds only in confusing the issue.

With all these words, we still haven't classified syllogisms — that is, we still haven't proved that the valid ones are valid and the invalid ones aren't. Since the derivations are all similar to one another, since it is good exercise (and sometimes even good fun) to carry out the derivations, and since the details are available in many sources (e.g., Hilbert-Ackermann), we shall not reproduce them here. We shall content ourselves by presenting one positive result, one negative one, and a statement of the facts.

We mention in passing that we have not described all the mnemonics above. The mnemonic words are sometimes arranged in a verse intended to make it easy to remember them and their function, and some invalid syllogisms are proscribed by technical terminology (undistributed middle) and rolling Latin incantations (*ex mere negativis* — or *particularibus* — *nihil sequitur*).

If non-existent distinctions are ignored, eight valid syllogisms remain; they are

$$\begin{array}{c} \mathrm{A\,A\,A} \\ \mathrm{E\,A\,E} \\ \mathrm{A\,I\,I} \\ \mathrm{E\,I\,O} \\ \mathrm{A^*\,O\,O} \\ \mathrm{A^*\,E\,E} \\ \mathrm{O\,A^*\,O} \\ \mathrm{I\,A^*\,I.} \end{array}$$

Here is a simple derivation (the one for A A A — Barbara); it uses the distributive and monotony laws for $\forall$, and the Boolean inequality

$$(p \Rightarrow q) \wedge (q \Rightarrow r) \leq (p \Rightarrow r):$$

$$\begin{aligned} \forall(p \Rightarrow q) \wedge \forall(q \Rightarrow r) &= \forall((p \Rightarrow q) \wedge (q \Rightarrow r)) \\ &\leq \forall(p \Rightarrow r). \end{aligned}$$

Here is a sample rejection (the one for A A*I — Darapti). The question is this: does $\mathrm{I}(p, r)$ belong to the filter generated by $\mathrm{A}(q, r)$ and $\mathrm{A}^*(p, q)$? If $p = q = 0$, then

$$\mathrm{A}(q, r) = \forall(q \Rightarrow r) = 1$$

and

$$\mathrm{A}^*(p, q) = A(q, p) = \forall(q \Rightarrow p) = 1,$$

but

$$\mathrm{I}(p, r) = \exists(p \wedge r) = 0.$$

Thus, the inequality

$$\mathrm{A}(q, r) \wedge \mathrm{A}^*(p, q) \leq \mathrm{I}(p, r)$$

fails when $p = q = 0$.

Before leaving syllogistics, monadic logic, and the field of logic altogether, we mention one half-philosophical half-mathematical difficulty that arises in the theory of syllogisms and indicate the way it is solved in the theory of monadic algebras. Some readers might find some psychological security in knowing that the concept exists and has a perfectly algebraic treatment.

Consider the most famous example of a syllogism, the one whose premises are "All men are mortal" and "Socrates is a man," and whose conclusion is "Socrates is mortal." In highly informal language, the trouble with this syllogism is that the algebraic theory, as we have seen it so far, is equipped with generalities only, and is unable to say anything concrete; Socrates, however, is a concrete individual entity. We have seen how a monadic algebra can be taught to say "All men are mortal." The reason is that "manhood" can easily be thought of as an element of a monadic algebra, since it is the obvious abstraction of the propositional function whose value at each x of some set is "x is a man." Socrates, on the other hand, is a "constant," and there is no immediately apparent way of pointing to him. (The use of the word "constant" in such contexts is quite different from its earlier use in the phrase "constant function." The important concept now is what logicians call an *individual constant*.) A classical artifice, designed to deal with just this difficulty, is to promote Socrates to a propositional function, that is, to identify him with the function whose value at x is "x is Socrates." This procedure is both intuitively and algebraically artificial; Socrates should not be a propositional function but a possible argument of such functions.

To find the proper algebraization of the concept of a constant, it is necessary only to recall that the elements of a monadic algebra are abstractions of the concept of propositional function, and to determine the algebraic effect of replacing the argument of such a function by some fixed element of its domain. If p is a propositional function with domain X, and if $x_0 \in X$, then $p(x_0)$ is a proposition, or, via the obvious identification convention, a propositional function with only one value — a constant propositional function. The mapping from propositional functions to constant propositional functions that is given by $p \mapsto p(x_0)$ (the evaluation map induced by x_0) clearly preserves the Boolean operations (that is, join, meet, and complementation). If the function p itself has only one value (equivalently, if $p = \exists q$ for some q), then that value is $p(x_0)$, that is, the mapping leaves the range of $\exists$ elementwise fixed. If, on the other hand, $\exists$ is applied to the (constant) function $p(x_0)$, the result is the same function, that is, $\exists$ leaves the range of the mapping elementwise fixed. These considerations motivate the following general definition: a *constant* of a monadic algebra $\mathbf{A}$ is a Boolean endomorphism c of $\mathbf{A}$ such that

$$c(\exists p) = \exists p \qquad \text{and} \qquad \exists c(p) = c(p)$$

for all p.

This definition is applicable to the mortality of Socrates. If, as before, q is manhood and r is mortality, and if Socrates is taken to be a constant, say c, of a monadic algebra containing q and r, then the algebraic justification of the classical syllogism is this: if **A** is the monadic algebra of a monadic logic such that both $\forall(q \Rightarrow r)$ and $c(q)$ belong to the filter of provable propositions, then $c(r)$ also belongs to that filter. For a proof, suppose that the filter **F** of provable propositions contains $\forall(q \Rightarrow r)$ and $c(q)$. Since

$$\forall(q \Rightarrow r) \leq (q \Rightarrow r),$$

and since c is a Boolean endomorphism that leaves the range of $\exists$ (and hence also the range of $\forall$) fixed, we have

$$\forall(q \Rightarrow r) = c(\forall(q \Rightarrow r)) \leq c(q \Rightarrow r) = (c(q) \Rightarrow c(r)).$$

Therefore, $c(q) \Rightarrow c(r)$ is in **F**, and hence so is $c(r)$.

Constants are much more important than their more or less casual introduction above might indicate; the concept of a constant (suitably generalized to the polyadic situation) is probably the most important single concept in algebraic logic. This should not be too surprising; in the intuitive interpretation, the constants of a theory constitute, after all, the subject matter that the propositions of the theory talk about. Algebraically, constants play a crucial role in the representation theory of monadic (and polyadic) algebras.

The theory of monadic algebras is an algebraic treatment of the logic of propositional functions of one argument, with Boolean operations and a single (existential) quantifier. In modern (predicate) logic and in virtually all of mathematics we need to use propositional functions with any finite number of arguments, and hence infinitely many quantifiers, one for each argument. The proper framework for an algebraic treatment of predicate logic is polyadic algebra. A study of this topic is beyond the scope of a short introductory book such as this. The interested reader can consult the book *Algebraic Logic* by the first-named author.

Index